M. J. GROVER

The Gold Prospector's Field Guide

A Modern Miner's Handbook for Successfully Finding Gold

Copyright © 2022 by M. J. Grover

All rights reserved. No part of this publication may be reproduced, stored or transmitted in any form or by any means, electronic, mechanical, photocopying, recording, scanning, or otherwise without written permission from the publisher. It is illegal to copy this book, post it to a website, or distribute it by any other means without permission.

All effort has been made to provide accurate information in this book. The author assumes no liability for accidents or injuries sustained by readers engaging in activities described here. All persons are advised that changes (man-made or natural) can occur at any time that may create hazardous conditions.

Laws can change at any time, and it is each individuals responsibility to be up to date with the latest regulations. Whilst we endeavor to keep this information up to date and accurate, any reliance you place on this material is done strictly at your own risk.

First edition

ISBN: 978-1-7362750-3-0

This book was professionally typeset on Reedsy.
Find out more at reedsy.com

Contents

I   Acknowledgement

Our Attraction to Gold	3
Valuable for Thousands of Years	4
Current Uses for Gold	7
Early Gold Rushes of the United States	9
Eastern States	10
Western States	16
Finding Historic Mining Locations	44
Researching Historical Mining Regions	45
How to Identify Historic Mining Sites	49
Geology of Gold	59
How Does Gold Form?	60
Basic Geology to Locate Gold Deposits	61
Gold vs. Pyrite – Identifying "Fool's Gold"	67
Mining Placer Deposits	69
Placer vs. Lode	70
Types of Placers	72
How Gold is Deposited in a Waterway	76
Where Can I Legally Prospect	79
Land Ownership	80
Mining Claims	84
Constantly Changing Mining Regulations	88
Gold Prospecting Equipment & Use	91
Basic Prospecting Kit	92

How to Pan for Gold	96
Sniping	100
Sluicing	102
Highbankers, Dredges, Trommels & Drywashers	105
Metal Detecting	115
Cleaning and Selling Your Gold	122
Getting Fine Gold Out of Black Sand	123
Separating Gold by Size	127
Gold in Rock – Crushing Ore	127
Methods of Cleaning Gold	130
Fair Prices Depending on Size and Quality	133
Selling Gold Dust, Nuggets and Specimens	133

Acknowledgement

This book is dedicated to the hard working miners over the centuries who left the comforts of everyday life to risk it all, and to those who still scratch at the Earth in search for that elusive yellow metal.

Our Attraction to Gold

Valuable for Thousands of Years

Gold is among the earliest types of metals to be extracted and used. It's brilliant yellow color has been valued for thousands of years and primary uses of gold were for ornamental arts and jewelry. In ancient civilizations all around the world, it was the wealthiest members of society who were adorned with the precious metal.

Its name originated from an old English word that referred to the color yellow, 'geolu'. Gold was used by craft persons of ancient civilization to make not only wearable jewelry items, but also for embellishing temples, tombs, and as ornaments for their kings and in making idols.

Gold has been linked to immortality, gods, wealth and rulers in various cultures worldwide. It is recognized worldwide as the most valuable metal, a belief that has remained for millennia and continues today. There is a universal recognition of gold's value.

Egypt

Gold was used by the ancient Egyptians as early as 3,000 BC, playing a very important role in ancient Egyptian mythology, which made it prized by royalty and priests. Not only did the ancient Egyptians create some of the world's most impressive artistic works of gold, they were responsible for the origin of many goldsmithing skills. One of these methods for casting gold is known as the "lost wax" technique, which is still used by jewelers today. This involves pouring the molten metal into a mold created from a wax model. After the mold is made the wax model is melted and drained away. Hence the term "lost wax." They also pioneered many other ways to use gold in decorations.

Gold's important place in ancient religions is shown by the large amount of gold that was used when constructing wealthy and powerful people's tombs.

Discussion of gold's prominence to the ancients was also described in the earliest texts.

Egyptian goldsmiths developed related skills such as the grading of gold's purity. They also made intricate maps to record the location of their gold mines, which fed their insatiable appetite for gold.

Greek & Roman Empires

Nearly all of the later ancient Mediterranean civilizations such as Greeks and Romans, prized gold and used it in all sorts of decorations, which required them to develop their own goldsmiths. Additionally, gold was particularly prominent in their own religions.

For the ancient Greeks, gold was not just considered to be a benefit to the dead but also a necessity. They believed departed souls could never cross the River Styx to reach the land of the dead unless they first paid the boatman. Naturally, he demanded a gold coin as the price of passage. This is where the custom of placing a coin in the mouth of the deceased came from.

Gold was used by the Romans as a setting for precious gemstones. The uses for the precious metal in ancient cultures were nearly endless. They included crowns, symbolic status, votive offerings, scepters and libation vessels. To be able to show one's status after their death, gold items were buried with the deceased.

South & Central America

The ancient tribes of the New World have been fascinated by gold for millennia. Most people are well aware of the value that the Aztec, Inca, and Maya cultures placed on gold and silver. Of course, a large part of the story relating to these cultures is the arrival of Spanish and Portuguese explorers who pilfered the countries of South and Central America to bring their golden treasures back to Europe. This dark part of our history is yet another example of the lust for gold that has dominated humans for thousands of years.

The Inca civilization in Peru believed that gold was the sweat of Inti (the Sun God) and because of this belief gold was used in manufacturing all sorts of religious items like sun discs and masks. In Colombia, gold was used in extravagant coronation ceremonies by covering the body of the future king using powdered gold. This created the legend of El Dorado.

Gold as Currency

Gold has also been used as a standard for currencies all over the world. Many ancient cultures used pure and alloyed gold coins as a medium of currency. Gold has notable qualities that would have made it desirable even back before there was a real monetary value. It was also highly malleable and tarnish resistant making it a perfect medium for coinage.

Due to the value people have always placed on gold, it became a form of barter for millennia before first being used to produce money. This led to gold being transported many thousands of miles from where it was first mined and smelted. The new owners either kept it as a treasure or melted the gold down to form their own artwork.

Some of this arrived as jewelry or raw gold and was melted down to create their own treasures. One of the early examples is gold jewelry which was found in the recovered sites of the very oldest human societies in Mesopotamia, the valley located between the Tigris and Euphrates Rivers in what is now Iraq. This part of the world has so many ancient finds it is often called "the cradle of civilization."

Current Uses for Gold

The early use of gold for ornamental purposes continues today. The primary use of gold since ancient days has been making jewelry, but approximately 75% of gold that is used now is still used in making of jewelry because of its beauty.

Gold can be melted and cast into very detailed shapes. The special properties of gold have allowed it to be used for various ornamental objects in the past and still today. Today, jewelers still use gold to make necklaces, rings, earrings and bracelets and pendants. In countries such as India gold is used to adorn the body, their belief is that by doing this they attract wealth and blessings. India is the largest user of gold for jewelry today.

In our contemporary society, gold is used by individuals for various applications beyond jewelry. It has a variety of special properties that make it a very useful metal. It conducts electricity extremely well, has a brilliant luster, and does not tarnish.

Gold is very simple to work with, and can be hammered into sheets and even small wires. It is extremely soft and malleable, and can be alloyed with various other metals to increase hardness and change its colors.

Gold has been and is still used as a valuable currency due to the fact that it is highly priced and its rarity makes even small pieces of gold valuable. And because of its universal recognition and value, it is very easy to liquidate gold to cash.

There are various means to invest in gold such as gold IRA, gold coins, gold bullion and gold bars. Mineral collectors often keep gold in its natural form as nuggets, dust, and specimens.

A growing amount of gold is used in making electronic devices. This is because gold is a very effective conductor that is capable of carrying very small electrical charges.

Gold has also proved to be useful in the field of medicine. Gold has been used by dentists for centuries to repair broken teeth with fillings and also replacing missing teeth.

Gold is also essential in the aerospace industry for lubricating mechanical parts, conducting electricity, and coating of space vehicles. There are literally thousands of electronic applications that gold can be used for, and as our technology advances in the future it is very likely that the demand will rise.

The combination of increasing cost of production from gold mines, as well as the increased wealth of many emerging countries ensures that the demand for gold will remain for the near and distant future alike. As new cultures increase their wealth, there is an ever growing demand for the precious metal.

The value of gold is likely to remain very high for the foreseeable future. Gold has been in demand for thousands of years, and that isn't going to change. This is good news for the adventurous gold prospector!

Early Gold Rushes of the United States

The first gold discoveries in the US were made on the East Coast. While they would later be dwarfed by the major gold strikes throughout the West, they caused much excitement for many decades. The United States would eventually become one of the richest gold producers in the world. It started with a young boy's discovery of a shiny rock in North Carolina, and would eventually lead to the massive gold rushes to California and Alaska.

Gold has now been found in nearly every state in the US. There are prospectors who find gold just about everywhere, but there are certainly better locations than others. The first step to finding gold is to learn about where to "old timers" found it.

This chapter covers some of this main gold discoveries throughout the US, and some of the specific areas where mining has taken place. But even with all this information, this chapter just scratches the surface. If you are serious about finding gold, then you need to make research a major part of your prospecting efforts. Knowing where to look for gold is just as important as learning how you find it.

I highly recommend that you seek out historical reports, books and other information for your specific area. There are entire books dedicated to specific states, and mining reports from various state and federal agencies that focus on specific mining districts. I have a bookshelf full of books for different areas. You can never have too much information.

Eastern States

Many people don't even realize how much gold was discovered on the East Coast. In fact, the earliest major discoveries in the US were found throughout the Appalachian Mountains, and there was a vibrant mining industry for several decades prior to the rush to California.

North Carolina

The first discovery of gold in the US is credited to a 12 year old boy by the name of Conrad Reed. He stumbled upon a huge gold nugget along the banks of Little Meadow Creek that ran across the family farm. His father John Reed organized some local men and established a small gold mining operation. They soon started finding more gold… a lot more gold! Big gold nuggets weighing several pounds, including a 28-pound nugget and many others weighing over a pound were recovered from the early mining activities.

As time passed the capacity of the mining operations began to increase. The mine was producing good gold and Reed continued bringing in more men and equipment. Some records showed that by the year 1824, $100,000 was realized from the gold captured using the pans and rockers on Little Meadow Creek.

In the early 1820s many more mines were established in the surrounding area. Eventually, hundreds of mines would spring up throughout North Carolina's western region. Miners began to realize that the gold deposits were more widespread than first imagined, and within a few years there were gold strikes being made all throughout the Appalachian Mountains.

Davidson County has numerous scattered lode deposits, and all waters have good potential for placer gold. The Uwharrie River and its tributaries are worthy of investigation, as placer gold has been found there. Montgomery, Stanly, Cabarrus, and Union Counties have plenty of historic gold production also. The Rocky River has gold and is worthy of prospecting. Other areas throughout the Carolina Slate Belt are all worth exploring.

Alabama

The Alabama goldfields were primarily found in a gold belt covering an area of 60 miles wide and 100 miles long. The gold belt enters and trends the northeastern part of the state coming from the border with Georgia towards an area in central Alabama known as the Piedmont Uplift.

The first major strike occurred in 1830 at Blue Creek and Chestnut Creek, and more gold discoveries continued throughout the coming years. Recovery of almost 80,000 ounces of gold was documented from 1830 to 1990. Gold has been found throughout Talladega, Tallapoosa, Chambers, Coosa, Clay, Chilton, Elmore, Cleburne, and Randolph Counties.

Tallapoosa County contains numerous gold districts including the Hog Mountain district, one of the biggest gold producers in the state. Much of the gold here was recovered through the cyanide leaching process, but creeks nearby produce placer gold. The Talladega National Forest has many creeks that contain placer gold as well.

In Clay County, placer gold can be found at Crooked Creek, Tallapoosa River, Wesobulga Creek, and many other streams throughout the county. Chilton County has gold in Coosa River, Blue Creek, Mulberry Creek and its tributaries, and Rocky Creek.

Numerous unnamed drainages will also produce placer gold for a hard working prospector. Some of the most valuable placers in Alabama are found in Cleburne County. Waters in the Chulafinnee Mining District will all produce gold.

Georgia

The celebrated county seat of Dahlonega in Lumpkin County was the center of the original gold strike in Georgia. For a time, this part of Georgia was getting more attention from gold miners than anywhere in the entire world. Miners scoured the rivers and creeks of Georgia in search of riches. The United States Mint even set up a branch in Dahlonega for several years. The gold that is mined in Georgia often assays at well above 23 karats, some of the purest natural gold deposits found anywhere on earth. As with other southeastern states, gold here can be found in both lode and placer deposits.

Reliable waters to try your hand in are the Chestatee River, the Tesnatee River and the Etowah River. Keep in mind that you should be watching for signs of old mining activity, because areas that were worked in the past are still going to produce for you today.

Cherokee County provides some pretty good placer opportunities, Two fairly rich lode mines were the Sixes Mine and the Cherokee mine, and the Little River and the Etowah would be worth your time and effort.

Check out the Nacoochee River and the Chattahoochee River in White County along with any smaller creeks. Historically, Dukes Creek has produced some really good-sized nuggets.

Placer gold has been found all through the Georgia Gold Belt, so don't just aim for the most well known spots. Dawkins, Forsythe, Paulding and Pickets Counties are all worth a look.

South Carolina

In South Carolina, the gold belt starts at Lancaster County to the north, and travels southwest toward Edgefield County.

Lancaster County contains one of the largest mines in the southeast, the Haile Mine. It is a lode mine that has produced over ¼ million ounces of gold. Numerous other gold mines are scattered throughout the county.

To the west, York County is also a good gold producer, with dozens of lode mines throughout the area. Many of the creeks and rivers in the area will hold placer gold. Check the Broad River and its tributaries, as well as the Little River in nearby Fairfield County.

Both Cherokee and Chesterfield counties have produced lots of gold over the years. Chesterfield County has had extensive gold mining since the early 1800s. For placer mining, based on the name, Nugget Creek might be a good place to start prospecting.

Saluda County lies within the Carolina Slate Belt, and has several gold mines and placer locations. Check the Little Saluda River and its tributaries. Edgefield County has a few productive lode mines near its border with Georgia.

Numerous other counties have potential for producing gold. Focus your efforts within the Carolina Slate Belt, as this is the area that contains the vast majority of productive gold mines throughout the state. Most commercial efforts here have been hard rock developments, but placer gold can be recovered by panning and sluicing.

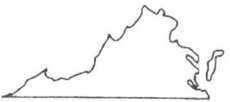

Virginia

The majority of the gold that has been mined along the eastern side of the Blue Ridge Mountains. A few small gold discoveries were found here as early as 1785, but small scale mining didn't start until around 1804. For approximately 25 years starting in 1828, Virginia was averaging a recorded volume of 3000 ounces of gold per year, with peak years reaching over 6000 ounces. Mining continued in full force until the famous discovery of gold in California, when thousands of miners pulled up and headed West. Mining in Virginia continued at only a small scale after that.

One of the best known gold mines in Virginia was the Franklin Mine in Fauquier County. There are numerous other lode mines that existed throughout the county. In fact, the counties in the northwest section of Virginia were some of the historically richest mining areas. Good gold was found throughout Fauquier, Culpeper and Spotsylvania Counties.

Continuing further south, several of the early mining camps were found within Buckingham, Fluvana and Louisa Counties. Again, the eastern flanks of the Blue Ridge Mountains around Charlottesville and Lynchburg are all auriferous. There were reports of multi-ounce gold nuggets being found in Tongue Quarter Creek, a tributary to the Willis River.

The **James River** flows right through some of the richest gold-bearing areas of the state, and there are still some decent placer deposits in this river. Rappidan River and Rappahannock River are worth exploring.

Other Minor Eastern Deposits

While most of the significant gold mining took place in the Alabama, Georgia, North Carolina, South Carolina and Virginia, there were much smaller gold occurrences found all across the East Coast. Most were not large enough to cause and sort of gold rush, but some did have enough gold to draw some attention from miners.

These were often glacial gold deposits, which occur when gravel is transported down from Canada by glaciers during past ice ages. These deposits can occasionally be rich, but their distribution is generally spotty and not rich enough to sustain large mining endeavors.

Minor gold occurrences have been mined in Maine, New Hampshire, Vermont, Maryland, Tennessee, Vermont and the states around the Great Lakes. Just about every state on the East Coast has *some gold*, but not enough to garner much interest from the early miners. Nowadays, the high price of gold and the interest from casual prospectors still brings folks out to search for those tiny golden specks in the local creeks.

Western States

The gold discoveries made in the West completely dwarfed those on the East Coast. The first major gold strike at Sutter's Mill along the American River would transform the country forever, bringing hundreds of thousands of men to venture West in search of fortune. The mining that started in California would eventually span out to cover every corner of the West.

California

The California Gold Rush that started in 1849 was by far the most significant gold discovery in the United States, and the state remains one of the top gold producers in the countries history. It attracted hundred of thousands of hopeful miners from all around the world. The gold deposits here absolutely dwarfed any gold discoveries that had been made prior.

The original discovery of gold at Sutter's Mill near present day Coloma was only the beginning. Miners would soon find extremely rich gold deposits throughout the Sierra Nevada Mountains. The American River, Feather River and Yuba River, along with literally thousands of smaller creeks and gulches were worked by small-scale miners during those first years.

The major river drainages have been prospected hard over the past century and a half, but high waters refresh placer deposits every year. There are literally thousands of small streams, gulches, bench deposits, and lode prospects throughout the Mother Lode country. The geologic conditions produced an amazing amount of lode gold deposits, which even today still hold great potential for prospectors. And it is these lode occurrences that feed the countless placers in the creeks, rivers, and gulches throughout the Mother Lode.

Although cost of production limits some of the gold mining in California today, it is really the political and environmental limitations that affect the gold output today. Much of the gold found today is by individual prospectors, but even small-scale methods like suction dredging are currently banned in the state.

Another extremely rich prospecting area in the state is the Northern California goldfield that covers much of Siskiyou, Shasta, and Trinity Counties. Gold nuggets were discovered here in 1851, and brought many prospectors from the Sierra Nevada Mountains north to the Trinity Mountains in search of gold. Good prospecting can be found around Redding, Weaverville, Yreka, and many of the smaller towns here. There are many remote areas here that still hold fantastic potential for gold.

Although largely overshadowed by more well-known gold discoveries in the northern half or the states, many parts of Southern California also produce gold. In fact, gold miners of Spanish and Mexican origin were mining this region prior to the more famous gold discoveries to the north.

The amount of gold deposits across the state of California is truly incredible. After the California Gold Rush, it didn't take long for men to start exploring every corner of the West in search of the next gold strike. Significant gold strikes have been made in every state in the Western US.

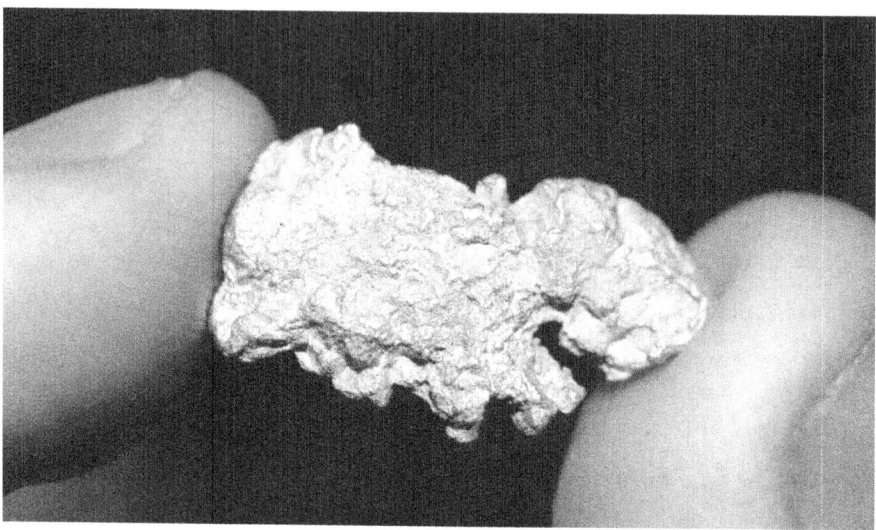

Arizona

Some of the first gold discoveries in Arizona were made along the lower Gila River near Yuma and along the Colorado River and the La Paz Mining District near present day Quartzsite. Additional mining areas sprang up in the desert around Quartzsite. The Dome District, or Gila City placers, is located at the northern end of the Gila Mountains east of Yuma. This area has been worked extensively since 1858, and most paying gravels have been reworked at least once. Many nuggets have come from the area, but distribution is known to be spotty.

In the years to follow, there would be several major gold rushes that would place Arizona as one of the leading gold producing states in the country. The Vulture Mine near Wickenburg holds the record as the most productive gold mine in the state. Rich placer deposits were also mined extensively in this area, particularly around Little San Domingo Wash.

The Bradshaw Mountains have been mined extensively since 1863 when Brothers Isaac and William Bradshaw separately discovered gold in the area. However, mining did not start until after 1873 due to the hostility of the local Apache Indians. The mountains are still home to some of the most productive gold placers in Arizona. Much of the gold here was recovered by small one-man drywashing operations. The dry gulches may show signs of past mining activity, but time can make them difficult to find.

Lynx Creek near Prescott is one of the richest areas of the state. It has been worked hard over the years, but gold can still be found. The majority of gold recovered from this area has been placer deposits. There is now a public panning access here that is popular with prospectors. Groom Creek is another productive drainage south of Prescott. Much gold has come from placers on the Hassayampa River and many lode mines.

Rich Hill has been one of the major producers in Arizona. Placers can be found at Weaver and Antelope Creeks. This area is very popular among metal detectorists, as it has a history of producing large nuggets.

Gold was discovered in the Greaterville in 1874. This created a small boom that brought in many miners which led to a thriving mining town. Water was always a challenge in this area. Ditches were dug to bring in water and work the placers.

Oatman is an old mining town located in the Black Mountains of Mohave County. Gold was discovered in the area in 1863 by Johnny Moss but mining did not start until early 1900 when new transport technology made it possible for miners and the gold to be transported. The richest gold ore was discovered in 1915 leading to one of the last gold rushes in Arizona. The last mine closed down in 1941 and by this time mines around Oatman had cumulatively produced gold worth approximately $2.6 billion in today's value.

The Gold Basin District is located 50 miles north of Kingman, 10 miles south of the Colorado River. Most gold comes from lode deposits. The Colorado River itself has very fine placer gold present.

There are hundreds of other gold-bearing areas scattered throughout Arizona. Yavapai, Mohave, Maricopa, La Paz, Pima and Yuma counties are the top producers. Arizona is one of the best states in the US for gold prospecting today.

Colorado

Prospectors headed for California's gold country in 1849 panned small amounts of gold along in the area around the South Platte River in Colorado. Despite the discovery, compared to visions of unlimited wealth coming out of the Mother Lode in California, little interest was given to these newly discovered placers and the gold seekers continued westward.

Nearly a decade passed until gold seekers exploring the region panned Cherry Creek near what is the present day city of Denver. Rich gold deposits were discovered and word quickly spread, attracting men from all around the country to the area. Although it was nearly 75 miles away, the visions of Pike's Peak on the horizon coined the name of the Pike's Peak Gold Rush, as prospectors would know they were close when the caught a glimpse of the massive peak.

As with most of the gold rushes across the west, thousands of men uprooted themselves and headed for the rich gold fields. An estimated 100,000 people originally left in search of Colorado gold. Travelers would paint "Pike's Peak or Bust" on the side of their wagons as the ventured across the country.

Miners quickly claimed up ground surrounding Cherry Creek, and ventured outward in search of more nearby gold. For the first several years of the rush, most mining was concentrated near the South Platte River. Within a short time, Denver City, Boulder City, and Golden City sprang up and became substantial mining communities, supply miners with all the picks, shovels, and whiskey they could drink.

West of Denver was the Central City Mining District has produced over 4 million ounces of gold, with the vast majority coming from lode deposits. The Argo Mill in Idaho Springs processed all the ores in the surrounding area and is now open for tours. The Clear Creek Canyon has produced a lot of gold and is a great area for prospecting now.

As prospectors radiated out in search of more gold, they quickly discovered new deposits to the southwest at the Breckenridge and Fairplay Districts. Further exploration revealed that rich lode deposits were present throughout the region, and these ores were mined extensively.

Later years would result in additional gold discoveries at Cripple Creek near the base of Pike's Peak, as well as additional discoveries at the Telluride and Summitville Districts in the southwestern part of Colorado.

The Leadville District is full of numerous mines, including many streams and gulches that were mined. The early miners were panning gold from the waterways, but heavy black sands clogging their sluice boxes ended up being a rich silver ore. This would lead to the development of silver mining in Leadville, which became one of the richest silver mining regions in the world.

There are numerous placer locations along the Arkansas River and several tributaries. Search for the telltale signs of historic placer workings. Gold nuggets have been found at a few locations in the headwaters, but most of the gold found on the Arkansas River as you venture farther east will be flour textured.

In general, you will need to focus on the fine gold deposits to have success prospecting in Colorado. Nearly all of the placer deposits across the state contain fine textured gold and small "pickers." Nuggets are quite rare. The Denver Museum of Nature and Science does have an amazing display of large nuggets and crystalline gold found in Colorado that you should definitely check out.

Idaho

A few minor gold rushes occurred in Idaho starting in 1860. Gold was first found at Orofino Creek near the town of Pierce. Another strike at Florence caused considerable interest in the winter of 1861, but the excitement was short-lived when the gold deposits were not as extensive as they had hoped.

Idaho's first major gold rush would occur in the Boise Basin, northeast of Boise. Within eight months of the strike, the area became the largest settled area of the Pacific Northwest. Boom towns jumped up everywhere with the four major ones being Idaho City (originally known as Bannock City), Placerville, Centerville, and Pioneerville (originally known as Dugem).

Silver was found on War Eagle Mountain in 1864, and thousands converged on the high desert valley of southwest Idaho to form the town of Silver City. Although, the vast majority of mineral recovered here were from hard rock deposits, gold was panned from Jordan Creek and a few other seasonal streams.

Gold is widespread throughout Idaho, and many rich gold strikes have been made throughout the state. The area around Elk City on the Clearwater, American and Red Rivers were all very productive. The richest deposits were found along the South Fork Clearwater River.

Other rich mining towns include Murray, Wallace, and Warren. The headwaters of the Boise River around Atlanta, Pine and Featherville have rich placers. Placers were worked extensively along the Snake River and Salmon River, though Wilderness designations now prohibit mining in many of these areas.

Oregon

Placer gold discoveries occurred in Southwest Oregon in 1851 when rich gold deposits were found at Josephine Creek and other nearby creeks. Soon additional gold deposits were found at the Illinois, Applegate, and Rogue Rivers, bringing hoards of miners from California and the Willamette Valley of Oregon.

Not long after the initial discoveries at Josephine Creek, rich gold deposits were found on the beaches at present day Gold Beach. Within weeks, thousands of gold prospectors were searching for gold in southwest Oregon.

The primary areas worth prospecting are in Josephine County, Jackson County, the southern end of Douglas County, and parts of Curry and Coos Counties. Gold here is found in both lode and placer deposits throughout the Siskiyou Mountains.

In Douglas County, check out the areas around Myrtle and Cow creek. Quines and Last Chance Creeks, along with the South Umpqua River and all waters that drain into it have been producers.

Jackson County is a great area to prospect. Well over ½ million ounces of gold have come from this county since its initial discovery in 1852. Check out the Applegate River, along with Sterling, Palmer, Willow, and Elk Creeks. Gold can be found in all waters in this area.

Josephine County has produced plenty of gold in the past. Check out the Illinois River and Josephine Creek, along with Galice and also Althouse Creek.

Basically all creeks in this county have the potential for gold. Check for old placers, hydraulic pits, and lode mines.

Curry and Coos Counties also have plenty of gold. Mining on the Sixes River will yield gold, and the Oregon beaches in this area also yield very fine gold. Look for black sand deposits.

Eastern Oregon was the other major gold producer in Oregon, with nearly 2/3 of the total production coming from a few counties in the northeastern section. A gold belt covers an area roughly 100 miles long and 50 miles wide, covering much of Baker, Grant, and Union Counties. The area is covered with old gold mines, hydraulic and hand placer areas, and valleys churned up by bucket-line dredges.

The eastern side of the gold belt starts along the Snake River next to the Idaho border. From Huntington up to the base of the Wallowa Mountains, several mining districts contain gold. The Burnt River has gold, and most major drainages that flow into it will too.

The Blue Mountains are a fantastic area for prospecting. The Powder River drainage upstream from the town of Baker has had extensive mining done since the early 1860s. Bucket dredges worked the Sumpter Valley for several years, and churned up miles of valley floor.

Other areas to check out include the old mining towns of Granite, Bourne, Greenhorn, and Susanville. The famous Armstrong Nugget was found in Buck Gulch in Susanville.

The gold belt continues westward toward John Day, Oregon. The John Day River and tributaries are all worthy of exploration. The majority of creeks throughout this area will produce placer gold, and thousands of mines and prospects still have gold across Eastern Oregon.

Montana

The first gold discoveries in Montana occurred in 1852, but it was the major strike on Grasshopper Creek in 1862 that brought gold seekers into the southwest corner of the state. The town of Bannack (west of present day Dillon, Montana) was built and soon thousands of miners were scouring the creeks and hillsides for the yellow metal. By the spring of the next year, huge gold deposits were found in Alder Gulch at Virginia City.

The creeks and rivers near the old towns of Bannack and Virginia City still have gold to be found, though much of it is off-limits today.

The Missouri River near Helena was the site of the Last Chance Gulch gold strike at the confluence of Orofino Gulch and Grizzly Gulch. More gold discoveries soon followed, and it wasn't long after that gold was found all throughout what is now part of the Helena National Forest. Nearly every drainage south of Helena was producing gold, and miners continued to flood in from around the world to explore the rich Montana mining grounds.

The headwaters of the Clark Fork near Butte produced gold. Much of the gold production in this region has come as a byproduct of the massive copper produced from this area. The Highland Centennial Nugget was found in the Highland Mountains south of Butte, and weighed over 25 ounces! It is currently on display at the Mineral Museum at Montana Tech.

Confederate Gulch is east of the Missouri River between Helena and Townsend. This was one of the richest placers in Montana history; some gravels reportedly paid over $1000 per pan! Nearby areas also include Boulder Creek, Cement

Gulch, Montana Gulch, and Montana Bar. Hydraulic operations were operated on much of the ground in this area.

The Ottawa District in western Montana produced rich placer deposits in Silver Creek, and many lodes have also been found since.

Rich lode deposits were found in the Georgetown District west of Anaconda.

A popular area for prospecting today is on Libby Creek, a tributary of the Kootenay River. There is a public panning area on Libby Creek about 20 miles south of Libby. The area is free of any mining claims, so the public can prospect here and keep the gold they find in the area. Just make sure you are within the designated area, as there are private claims adjacent to the panning area.

Far east of most of Montana's gold districts is the Kendall and North Moccasin Districts in Central Montana. Much of the gold here was in the form of low grade ores.

The area around Radersburg on the eastern flank of the Elkhorn Mountains has decent gold. Placer gold is present in many of the creeks in this area.

The Jardine District is located north of Yellowstone National Park. It has extensive hardrock deposits, and has also produced its fair share of placer gold as well.

The entire area around Cooke City, Montana has significant gold. Much of it can be found in small lode deposits and prospects that are scattered around the region.

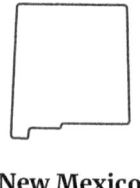

New Mexico

Much of the state's gold has come as a byproduct of mining for other minerals, but plenty of gold has been found in placer and bench deposits. The extremely dry climate in New Mexico has probably limited its gold production due to the lack of water. This also means that there is likely a lot of gold still available for the prospector to discover.

The Land of Enchantment was first prospected by its earliest inhabitants, the Mexican settlers and Spanish explorers. During these times, gold prospecting was made somewhat difficult by the native Apache tribes who were hostile to outsiders. It was in the mid-1800s that the early white explorers came to settle in New Mexico that prospecting for gold became more prevalent and gold production went widespread.

While a certain volume of New Mexico's gold came about as a byproduct of mining for other ores and minerals, there were certain areas that were regarded as rich and promising grounds for the supply of gold.

Since water is a limiting factor in this southwestern state, the best tools for finding gold in most areas will be drywashers and metal detectors. Searching in areas that gold has been found in the past is always a good idea. Here are a few of the most productive gold producing districts in New Mexico.

The Elizabethtown Baldy District was most productive along the west side of Mount Baldy, with paying gravels found in Grouse and Humbug Creeks, as well as the Moreno River. Well over 100,000 ounces came from this area. The Baldy placers occur on the east side of Mount Baldy, with rich gravels at South

Ponil, Ute, and Willow Creeks.

The Hillsboro District, located in the southwest part of the state, produced significant gold in cemented and uncemented gravels. Gold was being mined in several places around Hillsboro including Percha Creek, the Grayback Arroyo, Wicks Gulch and the Ready Pay Gulch among others. Many of these are dry placers with gold production coming from dryland dredges and drywashing in rich gulches in the area.

The Old Placers are located southwest of Santa Fe. Delores and Cunningham were productive gulches for small scale prospectors using drywashers. Nearby are the New Placers, which was known for producing very high purity gold.

Pinos Altos is a dry placer district just a few miles north of Silver City. Rich, Whiskey, and Santo Domingo gulches were all gold producers. Bear Creek produces placer gold.

The Rio Grande River is known to contain gold in many places. The geology of the area points to the Taos Mountains as the sources of the placer gold on the river. Gold deposits are also found in areas along the river and creeks draining in the river.

Several other districts have produced gold over the past two centuries. More detailed research shows gold throughout New Mexico in smaller deposits scattered throughout the state. The lack of water and extreme summer temperatures make prospecting difficult, but it also ensures that there is plenty of gold still left in the desert to be found by the hard working prospector.

Be aware that much of New Mexico is private lands, military reserves, and Indian Reservations that are either off limits, or require special permission. Many gold bearing areas on public lands are claimed, so be sure to seek out claim owners for mining access.

Nevada

Nevada is currently the #1 gold producing state in the U.S. While the initial discovery occurred when gold was found near Gold Canyon near Virginia City, the later discovery of low grade deposits in 1961 by the Newmont Mining Corporation are what solidified Nevada as the most mineral rich state in the U.S.

In the northern part of the state, Humboldt County has been a prospector's paradise for years. Gold can be found throughout the county, and it is well known for several rich areas that produce very large nuggets. Explore the Dutch Flats, Rebel Creek, Varyville and Winnemucca Districts, as all have produced several thousand ounces of gold. Using a metal detector around old drywash areas can be especially productive, as many gold nuggets were lost by the old timers using this method.

To the east in Elko County, numerous mining districts are found, each producing thousands of ounces of gold and silver. Most of the richest areas are in the northern part of the county, near the Idaho border and the surrounding the town of Mountain City. The Alder District near Wildhorse Reservoir was worked in the 1870s. In the Aura District, placer gold can be found in Sheridan and Columbia Creeks. The Charleston District is located in the Jarbidge Range, with gold mining on 76 Creek, Badger Creek, and along the Bruneau River. The Van Duzer District near the town of Mountain City produced significant amounts of gold, will several large gold nuggets being reported.

Pershing County is one of the best mining areas in the state. Known for placers often near the surface, a significant amount of gold has been found here. The early miners used drywashers to hunt for gold, but many prospectors today prefer to use metal detectors to locate "patches" of nuggets not yet found. The well known Rye Patch Placers is one of these areas, which has produced much gold since its initial discovery in 1938. Other productive placers include the Seven Troughs, Sawtooth, Placerites, and Rabbit Hole.

Dun Glen and Willow Creeks are also rich placer gold areas. Much of the recorded gold here comes from lode mines. Other districts worth exploring include the Sierra, Rochester, Humboldt and Unionville Districts. The overall richness of this county almost guarantees that plenty of gold is left in the ground awaiting discovery.

White Pine County has good gold prospecting opportunities, and like the rest of the state, its remoteness and harsh climate have limited its exploration. In the northwest part of the county is the Bald Mountain District. The Osceola District had seen extensive mining, including some hydraulic mining done in Dry Gulch. Several mines can be found in the viscinity of Ely, Nevada, which produce both lode and placer gold.

Nye County is one of the largest and most mineral rich counties in Nevada. Numerous gold districts are scattered throughout, and one of the largest gold mine in the U.S., the Round Mountain Mine is located here. Some notable districts include the Manhattan, Bullfrog, Jackson, Johnnie, Tonopah, Tybo, and Union. Numerous other districts can be found in Nye County, which have each produced thousands of ounces of gold as well as silver. This is a vast region with very rich mineral potential.

Mineral County, as you might expect from the name, has a rich mining history. The vast majority of gold from this area has come from lode deposits, and as a byproduct of silver mining. The Aurora, Candelaria, Garfield, Gold Range, and Hawthorne Districts were all producers. Lots of old-timer workings can be found in this county.

Lander County is home to the Battle Mountain District, where both lode and placer mining has produced significant gold. Bullion, Mud Springs, New Pass, and Reese River Districts are all worthy of investigation. Keep an eye out for silver ores.

In the southern corner of Nevada, both Clark and Lincoln Counties both have less mining activity than the central and northern parts of the state. Still, gold has been found here. Most has come as byproduct of large silver mines rather than from small scale placer mining. In Clark County, check out the Eldorado and Searchlight Districts. The Eagle Valley District had a small amount of placering done, and may be worthy of investigation.

Eureka County is home to the Carlin Mine, one of the richest gold mines in the U.S. The Buckhorn, Eureka, and Maggie Creek Districts are all worth investigation. Lynn District, home to the Carlin Mine, is found in the Tuscarora Mountains and many creeks nearby have been placer mined.

Esmeralda County has an abundance of prospecting opportunities. South of Tonopah is the Goldfield District, which has produced several million ounces of gold from lode deposits. Check out the Divide, Klondyke, Silver Peak, Sylvania, Tokop, and Tule Canyon areas. Much of the gold came from lodes, but some placers were worked by the Chinese as early as the 1870s.

Gold and silver are abundant throughout Nevada. Many of the smaller counties surrounding Reno and Carson City provide mining opportunities. In Storey County, the famous Comstock Lode produced fabulous amounts of both silver and gold. Lyon County is believed to be the home to the first gold discovery in Nevada, with continued productivity to this day. Douglas County has fewer gold deposits than its neighbors to the north, but gold can still be found here. In Washoe County, the Olinghouse District produced many thousand ounces of gold. Smaller prospects can be found throughout the county.

South Dakota

South Dakota has a rich history of gold production, starting with the initial discovery of gold at French Creek by a group of men led by General Custer into the area in 1874. The area known as the Black Hills quickly generated interest by men searching for their fortunes, despite the fact that the land was still owned by the Sioux Indians.

While several of the initial gold discoveries were small, rich placer deposits on Deadwood and Whitewood creeks confirmed that there were fortunes to be made in the Black Hills. An illegal settlement known as Deadwood quickly sprang up, one of the most lawless and dangerous towns in the country. Miners explored the area and found creeks filled with placer gold, as well as the rich lode sources.

As with most historic mining towns, thousands of men converged on the area and recovered most of the easily accessible gold by placer mining the rich creeks. After a few years, only a small percent remained, mostly opening up lode mining operations on a large scale.

The Homestake Mining company struck a rich ore and begun an expansion plan buying all the adjacent mines such as the Golden Star, the Crown Point, the General Ellison, the Gold Terra, the May Booth, the Old Abe, and the Giant mines. By 1900 the Homestake mining company owned over 300 claims, on a property estimated to be slightly over 2,000 acres and had over 2000 employees working in its mills.

The Homestake Mine went on to become the largest and deepest gold mine in the United States. It has an estimated production about 40 million troy ounces of gold over its lifetime.

There is still plenty of gold left to be found by gold prospectors today. The areas most worth checking out will be within the Black Hills in the far western part of the state, located near the border with Wyoming. Lawrence, Pennington, and Custer Counties have produced millions of ounces of gold. Although these areas have been extensively searched for nearly 150 years, there is still plenty left.

Surrounding the town of Lead, SD are numerous other lode mines as well as creeks and gulches that were worked. Deadwood, Strawberry, and Elk creeks all produced placer gold. Yellow creek was also very rich. Annie and Squaw creeks also produced, as well as numerous small lode deposits in those areas.

At the southern end of the Black Hills in Pennington County, overall gold output has been much less than in Lawrence County, yet many thousands of ounces have still come from here. Battle, Castle, Rapid and Spring creeks will all produce placer gold, with numerous active and abandoned lode mines still present in this area.

In general, the gold deposits in South Dakota are heavily concentrated in the above mentioned areas within the Black Hills. The central and eastern sections of South Dakota are generally devoid of gold. To be successful here, it will be especially important that you focus your efforts in areas known for historic gold production.

Utah

While Utah does not have extensive gold prospecting opportunities like many other western states, there are still plenty of places that you can find gold though. You just have to do diligent research and realize that there aren't as many places to mine compared to neighboring states.

Surprisingly to some, Utah is actually one of the top gold producing states, but most of this gold comes as a byproduct of silver, lead, copper, and zinc mining. In fact, the largest mine in the United States is located right here in Utah.

The Bingham Canyon Copper Mine near Salt Lake City is by far the largest producer in the state. The first claim on the canyon was the West Jordan claim staked in 1863, followed by the Vidette claim. Initially it was gold and silver where the main minerals mined in the area, until 1873 when a railroad was built that copper and other minerals were considered for mining.

As the world's largest excavation, the Kennecott Copper Mine has been an extremely productive copper mine. It was the second largest copper producer in the United States in 2010 producing over 18 million tons of copper. Gold and silver is a by-product, but due to the large scale mining the overall output of gold is also massive. The mine is still active.

A handful of other large commercial mines also produce significant amounts of gold as well, resulting in significant overall production for the state. These mines are responsible for 99% of the total gold that gets recovered an an annual basis.

Most placer discoveries throughout history in Utah resulted in minimal recoveries, as much of the gold was extremely fine, and not in significant quantities to warrant much attention from the early miners. Much more attention was given to other minerals.

There are thousands of old hard rock mines and prospects throughout Utah that produced some gold as byproducts of silver, lead, zinc, etc. Creeks and rivers near these mines always have potential to produce some gold for a placer miner, but many have had little reported about them due to lack of any significant concentrations. A few areas that have documented placer workings are listed below.

The general area southwest of Salt Lake City has fine gold in the creeks near the Bingham Mine in the Oquirrh Mountains. In addition to lode mining, many placers have been exploited in this area. Johnson Creek and Recapture Creeks in the Abajo Mountains have gold throughout.

Gold has been found in the Colorado River, but it is extremely fine and requires great care to recover it. A man could probably make his fortune from all the gold in this river, but each speck is so tiny that they are worth a fraction of a cent. It takes a lot of tiny specks of gold dust to amount to much money.

In the northeast corner of the state, the Green River has produced fine placer gold below Flaming Gorge Reservoir.

Several creeks draining the Henry Mountains in Southeastern Utah will produce some fine gold. The same can be said for the San Juan River in the southwest part of the state.

Utah is one of the poorer states in the West for prospecting. While huge amounts of gold have been mined from the Bingham Mine, the geology in most of the state is not productive for gold. To find it, you will want to focus your attention on the known areas.

Washington

Washington is a great state with plenty of opportunity for gold prospecting. Overall production has been much less than most of the other western states, but several million ounces of gold have still been found here. With a few exceptions, most gold here is quite fine, and distributed throughout the creeks and rivers.

There is gold in the Columbia River. It is the largest drainage in the Pacific Northwest, and drains all of Washington's waters east of the Cascade Mountains. Its headwaters start in Canada, and ends at the Pacific Ocean. Placer deposits are throughout the gravels of this river system, generally in the form of very fine flour gold. The same is true of the Snake River which drains into the Columbia in the southeast part of the state.

All of the counties in Washington have produced gold. East of the Cascades, the most noteworthy are Stevens, Ferry, Okanogan, Chelan, and Kittitas Counties.

Stevens County is in the far Northeast corner of the state. Hundreds of mine sites are scattered throughout the area, often lode mines that also produced copper, iron, etc.

Ferry County is one of Washington biggest gold producers, containing the Republic District 25 miles below the Canadian border. The vast majority of the gold found has been from lode mining. Placers produce mainly fine gold.

Okanogan County is in the northern part of Washington on the east slopes of the Cascades. The Columbia, Methows, and Okanogan Rivers all contain gold.

Chelan County was the largest gold producer in the state. Much of the gold came from lode mines in the Blewett, Chelan Lake and Wenatchee Districts, but numerous placer operations have also been in place since the mid 1800s.

In Kittitas County, you will find one of the best known prospecting areas in Washington. Near the historic town of Liberty, there are numerous lode and placer mines worthy of investigation. Williams and Swauk are the best producers. They are also well known for the unique wire gold specimens that have been found.

West of the Cascade Mountains there is also ample opportunity for gold prospecting. All counties have produced at least some gold in the past. A few of the more noteworthy include Whatcom, Skagit, Snohomish, King, and Pierce.

Gold can be found along the Skagit River. Also explore the smaller creeks and tributaries of the river.

Snohomish County has produced a lot of gold, mainly in the Monte Cristo and Silverton Districts. King County is also a large gold producer in the state, but nearly all sizable operations have been lode mines. Exploring creeks near these lode areas may be productive for fine gold.

All counties along the Columbia as well as the ocean beaches have fine gold. Look for black sand deposits and carefully sample the sands. Gold will be extremely fine textured.

Wyoming

Gold prospecting in Wyoming gets surprisingly little attention in comparison to some of its neighboring states, despite having vast mineral wealth. There are less people per square mile in Wyoming than any other state, so it is no surprise that it has been explored less than many other areas. Most think of Wyoming in terms of cattle ranching, oil, and natural gas, but plenty of gold has been discovered here too, although not nearly as much as many other western states. Long and harsh winters can make for a very short mining season.

The first reports of gold in Wyoming were as early as 1842, by westward travelers along the old emigrant trail at the Sweetwater River. Indian hostilities made prospecting efforts difficult for several decades. By the 1860s, prospecting efforts increased and several rich discoveries were soon found throughout the state.

This Sweetwater discovery lead to small groups of armed prospectors attempting to mine gold along the river, but this was a limited and dangerous endeavor due to the Indian raids.

By the late 1860s to early 1870s, the gold from Sweetwater River had been traced upstream to the Carissa lode. This event prompted Wyoming's first true gold rush and the creation of several mining districts.

The South Pass-Atlantic City District lies at the southern end of the Wind River Range. Additional districts in the area include the Lewiston, Twin Creek, and Oregon Buttes districts. Significant lode and placer gold has come from these areas. Production records were extremely poor from these areas during the early gold rushes here, but hundreds of thousands of ounces are known to have come from these districts, and quite possibly several times that amount.

In the central part of the state in the Granite Mountain Range, the Tin Cup, Rattlesnake Hills, and Seminoe Districts have been mined. Most exploration efforts have shown low grade gold ores, but limited investigations show potential for areas worthy of investigation by the recreational prospector.

The Absaroka Mountains in the northwestern part of the state produce gold in several areas. The Sunlight, New World, Kirwin and Stinking Water districts all have placer gold deposits that can be recovered by panning, sluicing and dredging. Limited commercial efforts have been put into this area, but recreational prospectors can find plenty of gold here.

The Laramie Mountain Range has several districts worth investigation. The Silver Crown, Garrett, Warbonnet, and Esterbrook all produce gold.

The Medicine Bow and Sierra Madre Mountains in south-central Wyoming were prospected extensively during the late 1800s up through the great depression. Numerous mines are scattered throughout the region and still have good potential to yield gold. Keep in mind that often these mines were abandoned because of low metal prices, not because they were worked out. With today's record high gold prices, there is a good chance that these old prospects are once again profitable for gold mining.

Wyoming has vast expanses of public land that are open for prospectors to explore. While the gold deposits are not as abundant as other states, the potential for new discoveries are good simply due to the overall lack of exploration in this sparsely populated state.

Alaska

There have been some huge gold nuggets found in Alaska. It is one of the most remote mining regions on Earth, and it still holds great potential for gold mining. There are vast unexcavated areas containing rich gold veins and ores remain unexplored which when opened up could push the mining operations well into the next century.

Alaska was the last state that gold was discovered in the United States. It was the famous gold rush to the Klondike region that brought the first white men to this remote part of the country. Prior to this discovery, only a small population of native people inhabited this region. At this moment, Alaska produces more gold than any of the other states of the United States except for Nevada.

The state of Alaska is rife with smaller placer mines with many miners continuing the enterprise over several generations. Rich gold mining districts are found in virtually all parts of the state.

There are many rich gold areas on the Seward Peninsula. The best known are the famous beach places of Nome. While most of the gold found in the sand at Nome is fine textured, nearby Anvil Creek just west of town has produced some massive gold nuggets over the years! Gold was first discovered around these parts in 1898 by the "Three Lucky Swedes" and since then over 3.6 million ounces of gold has been recovered from the area.

There are also many rich mining districts a few hundred miles southeast of the Seward Peninsula along the Yukon and Kuskokwim Rivers. The Kuskokswim

Gold Belt is described as broken belt of gold bearing rocks that have the reputation of producing large nuggets and considerable gold.

Some of the better known mining districts in this belt are the Fortymile and Iditarod districts. Two areas in this region that are well-known for producing large gold nuggets are Ganes Creek and Moore Creek, both of which were once popular destinations for gold prospectors using metal detectors.

Chicken Creek is another gold rich destination in this area where gold was discovered way back in 1896. Gold panning and placer mining is popular around Chicken. Recreational miners have reported finding multi-ounce nuggets in this area with nice nuggets continuing to be found.

Jack Wade Creek Gold Panning Area runs along the Taylor Highway just a few miles north of Chicken. There are several miles of creek that are unclaimed and open to panning, sluicing and metal detecting. This is another area that can produce some really nice nuggets if you put in the time.

Further inland, many gold bearing areas are found on the Yukon traveling upriver toward the Klondike. Rich placers are found around Fairbanks in the Yukon and Tanana River basins and on the Chena River. The gold discoveries near Fairbanks were discovered as late as 1902, and are considered to be the last great gold rush in America.

Plenty of gold has been found in the area around Cook Inlet in South-central Alaska. The Kenai Peninsula, Cache Creek, Willow Creek, and Valdez Creek are some of major gold districts around Cook Inlet and Anchorage. A lot of big nuggets have been found here too.

Alaska has very favorable mining laws when compared to all of the other states in the U.S. Restrictive regulations make it very difficult to operate large-scale mining operations in the lower 48 states, but Alaska still has decent laws that support the mining industry here. Most federal and state owned lands are open for recreational mining as long as regulations related to mining techniques are followed.

In most parts of the U.S. it is the lode mines produce the vast majority of the gold, but in Alaska there are still rich placers that have yet to be mined. Since tourism is such a large industry in Alaska, quite a few old family owned mines and placer mines have now been opened to the general public where gold panning and sluicing is allowed.

$1,250,000 of gold bullion mined from the beaches of Nome. At today's prices, this is nearly $100 million dollars worth of gold. Miners & Merchant's Bank, June 10th, 1906.

* * *

Finding Historic Mining Locations

Earlier we took a look at the some of the major gold discoveries made throughout the US. Now we're going to go even deeper and learn how to do in-depth research to find the exact locations where you are going to have the best chance of finding gold.

There is no one "right way" to find gold. However, there is no doubt that learning to identify historic mining sites, and being able to use this skill to find places where you can still search for gold, is critical for your success as a modern-day gold prospector.

Researching Historical Mining Regions

Gold prices have reached record highs in recent years, and there has been lots of interest by beginners who want to try their hand at gold prospecting for the first time. It is very possible to be successful if you do your research and learn how and where to find gold.

To start with, you need to find an area that has good potential for producing gold. It is very common for people who are just starting out to ask a question like, "There is a creek by my house, do you think there might be gold there?" Sure there might be, but this is really the wrong question to ask.

First, what you should be finding out is if gold has already been discovered there. This is the single most important thing that will determine whether a beginning prospector is successful or not. While there is a slim possibility that you might get lucky and find a new gold strike in an undiscovered location, you are going to have much better odds if you search in an area that is known for producing gold in the past.

Research is very important and will save you a ton of time and effort. In the mid 1800s, there were literally thousands of men scouring the countryside in search of gold. When a large amount of gold was found, a gold strike would occur and those men would converge into the area to mine for gold. These areas are well-documented. Take the time to find them and you have much better odds than if you just venture out blindly in search of gold.

Location-Specific Books

There are numerous books available about gold prospecting in various states that will give you some great information on where to start looking for gold. This is the best money you will ever spend. A location-specific book can provide the kind of information that you may never be able to find otherwise. Most of the major gold-bearing areas are very well documented, and general information is readily available if you look for it.

Government Reports

Books about prospecting in different states are excellent resources to get you started in your search for gold. They generally provide good information on specific rivers, creeks, and large gold mines that historically operated within a gold district.

Once you have identified a general area that has the potential for gold, it is time to start getting more in-depth in your research. This is where many prospectors stop, but reading books should just be the start of your research into a gold bearing area.

Some of the best information on gold occurrences is found in a variety of government reports. Many of these reports are now digitally available online, or can be purchased as reprints from a variety of sellers. Many different agencies produced various reports on mining throughout history, but they are most commonly put out by the U.S. Geological Survey, Bureau of Land Management, or the various state level agencies that manage mining and mineral resources.

Be sure to check out the 'Free Mining Reports' chapter at the end of this book.

Old Newspapers

This is one type of research that is overlooked by most prospectors. It may seem that finding information about gold mines in a newspaper would be like finding a needle in a haystack, but historical newspapers from the appropriate time period are actually full of great tips and pointers that can lead you to areas to search for gold.

Newspapers back in the 1800s were different than they are today. The bulk of the reporting was based around the economy at the time. In most areas of the U.S. during that time period, the reporting was based around farming, ranching, and mining.

Find old newspaper archives for the area that you are researching and you are sure to find some great tips on finding some great gold bearing areas. If the area produced some large gold nuggets, it was not uncommon to see news articles about the major discoveries.

Visit the Library

Want to dig deep and find some old historical information about a specific gold area? There is no better way to do this than take a trip to your local library! Most librarians really enjoy helping people with research. They are great at "digging deep" and finding old, obscure materials that contain great tidbits of information about mining history. Libraries can generally help you find the old historic newspapers that we discussed earlier.

You cannot have too much information on an area. If you are serious about finding gold there, than spend some time at home researching the various mines and prospects in the area. Find out where the richest mines were and consider where you might have the best chances of success the next time you head out.

Aerial Imagery

Many successful gold prospectors today use some type of mapping program like Google Earth to locate potential prospecting areas. This is an excellent tool to help you find areas, even if you are thousands of miles away in the comfort of your living room.

Surface disturbances like dredge tailings and hydraulic washouts (which we will discuss later) will show up clearly in these aerial photos, and you can use this in conjunction with maps and GPS to plot areas to investigate when you get there. You can also identify areas that would concentrate placer gold like bends in the river or large exposed boulders.

Just about every successful prospector I know uses Google Earth and other mapping software to research gold areas. The combination of aerial photos and topographical maps in combination with documented historical gold locations gives today's prospector an advantage that the early gold miners never had.

Sometimes the signs of historic mining are obvious; bucket line dredge tailings, hydraulic mining, and open adits are good examples of things to look for. Other locations you should look for include hand-placered streams, hand-stacked rocks, exposed bedrock and drywash piles. These are all indications of past gold mining. Finding these areas can be easy or difficult depending on your experience, but a bit of knowledge will help you train your eye to identify what you are looking for.

Books, maps, historical documents, aerial imagery, and everything else available to us give today's prospector a huge advantage compared to the early miners. If you are just starting out, I highly recommend that you use all of these resources that are now available, as they will greatly increase your odds of success in gold mining.

How to Identify Historic Mining Sites

Understanding the historic mining methods that were used can help us determine the likelihood of how and where the gold will be today. Many of these methods have not been used since the days of the early gold rushes, yet you can still see evidence of their past use on the landscape. Rest assured there is still gold in many of these locations.

Hydraulic Mines

Hydraulic mining was a gold recovery method that was used during many of the gold rushes around the world during the 1800s. It was used extensively in California's Mother Lode Country during the famous gold rush there.

One of the problems that the early California gold miners faced with basic placer mining was the amount of manual labor that was required to process the gravels. Most of the early placer mining was done by individuals or small groups of men who would shovel material and process it through sluice boxes or other equipment to extract the placer gold. If the prospector was fortunate enough to find a rich pay streak of gold then it was often profitable to work these gravels by hand. Thousands of men tried to strike it rich this way, and sometimes they were successful. However, most of the richest and most easily accessible gravels had been exhausted within just a few short years, and in order to mine profitably, it was necessary to process much more material than could be done with simple hand tools.

Hydraulic mining was the perfect answer to this problem. It used water under high pressure to wash away gravels and high bench deposits, making a slurry that could be run through a sluice box or other equipment to recover the gold. The water was routed from a source such as a nearby river or stream into ditches and flumes, where it was run through a canvas hose to a high pressure nozzle called a monitor. The monitors were huge cannons, some up to 18 feet long, capable of discharging massive amounts of water under extremely high pressure. These jets of water could literally wash away mountains, removing all the material down to bedrock. They were the perfect tool for working the ancient river gravels buried underneath mountains.

Hydraulic mining was developed by the ancient Romans. The process was adapted and perfected by gold miners in the 1800s, who would use high pressure water jets, called monitors, to wash away hillsides to release gold. Their effectiveness allowed miners to work areas that would have been unprofitable by other means.

Due to its large scale, hydraulic mining was a popular method used by larger mining companies, rather than individual prospectors. These operations would often employ dozens of men. They were extremely efficient at moving tons of material for significantly less cost than other methods. It made many areas with lower paying gravels profitable to mine where it may not have otherwise been cost effective.

Unfortunately, the downside of hydraulic mining was that it dumped massive amounts of discharge material into the nearby streams and rivers, choking them with debris and causing extensive flooding and erosion issues. The miners had no interest in stopping the practice, as it was a profitable venture for them. Farmers cried foul, as their fertile farmlands were covered with silt from the mining upstream. With the huge silt discharges into the river, flooding became a bigger and bigger problem, destroying not only rich farmlands, but towns as well. The war over hydraulic mining continued throughout the 1870s and 1880s. Farmers feuded with the miners, politicians got involved, and eventually the practice mostly ended in 1884.

The scale of these old hydraulic workings can vary; some are relatively small and were used on small creeks, while many of these areas are massive and can be seen from miles away. Gold can be found anywhere in hydraulic pits, but some investigation of the area can help you determine the best places to look. If you study the terrain, you can often find the areas where the monitors were set up. There is often an elevated piece of ground, and quite often it will be littered with lots of old iron rubbish. You can visualize the area where the water was washing away material, and locate the exposed bedrock and virgin banks that were just out of reach of the high pressure water.

Due to the high volume of runoff produced at these sites that would commonly overload the sluices, hydraulic mine sites were notoriously inefficient at times. There are still plenty of gold nuggets that were missed and still left to be found.

Lode Mines

Many of the largest mining operations were lode mines, which were the site of the hard rock gold deposits that usually fed the placers below. These mines are generally easy to locate, are often marked on maps, and have the telltale piles of waste rock and crushed ore lying around. Tunnels can often be located, although they are frequently caved in if they have not been actively worked for a long time.

These lode mines can be great areas to search for hard rock gold and specimens still locked up in quartz. Just be aware that some of these old mines are still in operation or located on patented claims so they may not be open to prospecting.

Gold mining underground was no easy task. It was dark, cramped and dangerous. The days were long and men worked every day of the week. Most were working for wages, so even if they made a rich strike, the riches belonged to the mine, not the miner.

Old lode mines can also be extremely dangerous. They may be structurally unsound, have poisonous gases, nasty chemicals, dynamite, bats, spiders, snakes, and all kinds of nasty stuff. Unless you have the proper training, you should never enter an old lode mine.

Instead, explore areas surrounding the existing mine. There are commonly nearby gold veins that were undiscovered or unworked by miners in the past.

If it was a large mine, there may have been a stamp mill on site to crush ore. Stamp mills during the early gold rush days were generally powered by water, although sometimes steam engines were used as a power source. Their construction typically involves a series of heavy metal stamps arranged in a wooden frame called a stamp battery. A system of rotating shaft and cam is used to raise and lower the stamps on top of the ore. Stamps are built heavy, made from steel or cast iron heavy enough to pulverize the ore beneath. The stamps are repeatedly raised and dropped onto ore that is fed into the mill, until the coarse ore is reduced to finer material capable of further processing.

The stamp mill provided an invaluable service for the mines, but not all mines had a stamp mill on site. Smaller mines often needed to transport the ore long distances to have it processed. Often this resulted in scattered piles of ore that would get high-graded and left behind. Mines would close, miners would move on, and rich ores would get left behind. It's not uncommon to find ore piles still sitting at old mine sites that have been left untouched for over a century.

Dredge Tailings

Not to be confused with the suction dredge used by many gold prospectors today, a bucket line dredge is a giant gold processing machine that can move literally tons of material in a short amount of time. They were used extensively throughout the world during the early gold rushes in the U.S. Today, most of them are out of commission and lay dormant. While they were extremely efficient and cost effective to operate, they caused irreparable damage to riparian areas. Today most are just relics of the past in various stages of decay.

The basic design of a bucket line dredge is relatively simple. The front of the dredge is comprised of a series of huge metal buckets called the boom. Picture a series of metal scoops about the size of a bathtub, designed to go around in a continuous loop similar to a huge chainsaw. The boom is angled down into the earth and digs up a continuous supply of gravels to be processed. The material was brought inside the dredge for processing and gold was separated from the worthless gravel.

The signs of past dredging are easily identified by the large rock piles left in relatively uniform patterns and rows. Viewed from aerial photos, it is easy to see the zigzag pattern that was left behind. Often there are small ponds throughout the tailing piles as well.

Metal detectors can be used effectively in tailing piles, but they are notoriously trashy and it can be difficult to locate gold without dealing with thousands of ferrous iron targets. They are very challenging to to hunt, but a patient prospector may be rewarded for his effort.

One exciting fact about bucket line dredges that might interest the modern-day prospector is the fact that they actually missed nearly all of the larger gold. Part of a dredge's design provided a way to separate the gold from the rocks and gravel by removing any material larger than 1/2" or so. This might seem counter-intuitive at first, since that means any large chunks of gold are separated out and rejected by the dredge. There was good reason for it though, as almost all of the gold in any given area is going to be small. Yes, a few nice nuggets are going to get lost, but the benefit of not having to process tons and tons of extra material was worthwhile. The result is large gold nuggets are still buried in the tailing piles waiting to be found!

The piles of processed gravel and rocks from these old dredges can still hold a lot of gold. Using metal detectors can work very well, but it takes a lot of patience and persistence.

Hand Workings

There are thousands and thousands of miles of creeks throughout the U.S. and the world that have been hand placered. This simply means that the gold was extracted without any specialized equipment. The miners simply used picks and shovels to get down to bedrock and find the gold.

Beginner prospectors might confuse hand placered creeks with the larger dredge tailings. The easiest way to tell the difference is that hand worked areas are usually less uniform in appearance, with stacks of rocks and overburden scattered around in random locations. Dredges required a large amount of water to operate, and were better suited for river systems with deep bedrock. Most small creeks and gulches were worked by hand rather than with dredges, which were generally used in the main river channel.

Creeks that have been placer mined by hand can still hold excellent potential for gold. Thousands of early miners recovered gold with nothing more than basic hand tools, and some were more efficient than others. Often they missed gold that was down in the cracks of bedrock and difficult to expose. Gold was also lost when they shoveled overburden into waste piles on their way down to bedrock. Sometimes there was gold in this material, and it is still waiting to be found today.

Sometimes miners would cut trenches into bedrock that would act as a natural sluice where gold would be captured. The miners needed some way to separate out the gold from the tons of worthless rock. Nowadays, prospectors can use simple equipment like sluice boxes to run the concentrates through, but the early miners didn't have the luxury of using the small, portable sluice boxes that we have today. Thus, they would employ ground sluicing.

The use of ground sluicing was completely dependent on water. Each site is different. If a miner was fortunate to have a creek on location, then diversion of water was relatively simple. In other cases, water was transported to the mining site through a series of ditches and flumes.

At many locations, miners would construct small reservoirs to capture the limited water that was available. Water would be collected and then used sparingly at select times to maximize its efficiency.

Captured water could be used a few different ways depending on the topography of the mining site. In some cases, water was diverted and released so that it would wash over gold bearing gravel, which would erode away the gravel and move it through the ground cut sluices at the same time. In other situations, the water would be run over a cut bank creating a waterfall. This would create additional erosion and help the process of breaking down the material above bedrock.

Each location will be different, and the ability to capture and retain gold varied from each site. Be on the lookout for small reservoirs, ditches, stacks of rocks, trenches cut into bedrock, and any other indicators that may look like old hand workings. The hand of time can often hide these sites, so keep a sharp eye out for these indicators. They can lead you to long forgotten mining sites.

Drywash Piles/Hillside Prospects

Many of the sites that gold prospects around the west were relatively insignificant. Often they were small outcrops where a prospector did a bit of digging to recover some exposed gold, but stopped after the gold ran out. In arid climates like the desert southwest, prospectors often used drywashers to prospect these locations, leaving behind waste material. The inefficiency of drywashers meant that quite a bit of gold was lost, and remains to be found in those old piles. Sometimes these prospects are located just uphill from a creek or river, and prospectors would pack the material down to the creeks.

Two miners drywashing for gold in San Bernardino County, California in the early 1900s. These old drywashers weren't very efficient, and they would lose a fair amount of gold. Large nuggets would often get screened off and fall into the waste piles. A modern-day prospector armed with a sensitive metal detector can often recover some nice nuggets from these old workings.

Sometimes seemingly small and insignificant hillside prospects are rich with gold. The old timers did not have the efficient mining equipment that we have today, and they missed a lot of gold. Metal detectors can be a great way to explore these areas, and find gold nuggets that the old timers missed.

Locating old mining areas is a great way to find gold today. While it is certainly possible to locate patches of gold in virgin ground, finding these sites where gold has been mined before is by far the easiest way to recover gold. Identifying the type of mining activity that took place can help you determine where you are most likely to find the gold that was missed, and how you can best add some gold to your poke today.

Geology of Gold

How Does Gold Form?

While diamonds can be artificially produced, there is only so much gold available on the planet. It cannot be created in a lab the way that many gemstones can be. While some scientists believe that gold mining in outer space may be the bonanza of the future, such dreams are currently too expensive to be realized. Of course, if the value of gold continues to climb up and up, maybe one day soon we will be exploring meteorites for precious metals. For us small-scale miners interested in gold, the search remains near the surface of our world where we find it in hard rock and placer deposits.

Water not only moves gold by erosion, but it actually helped form the gold veins that originally formed many millions of years ago. Underground sources of hot water and pressure combined to melt gold and sulfur, which are often found together, and push them towards the surface. When the waters cooled, gold filled in the natural cracks in the rocks which formed veins of gold. These veins are often microscopic, but sometimes they are quite significant high grade deposits that can be mined profitably.

These processes mentioned above tend to occur in areas with natural faults that move the earth's crust. This is the reason why the areas that contain the most gold are usuallybmountainous regions with constant geological activity going on.

Although gold can potentially be found anywhere on earth, the truth is that most gold is found in the more active parts of our planet, geologically speaking. Most of the original discoveries of gold were made by accident around stream beds where gold particles and nuggets were found washed down from the hills and mountains where most of the gold veins are located. These placer deposits are where the majority of present day gold miners go looking for gold, as these concentrations can be some of the most productive areas. The erosion process of water over time reveals the places in which large gold deposits can be found. Streams that cut through the rocks carry the smaller elements of gold down with them.

Because of the malleable nature of gold, it's rarely found in the form of nicely shaped crystals. Raw gold is mostly found in lumps or irregular masses that filled the cracks and natural faults in the rocks. These are the typical gold nuggets that we have come to associate with natural gold.

On rare occasions however, gold can be found in crystalline forms, which are very rare and highly collectible. Gold specimens of rough texture that contain quartz and other host material are also occasionally found by prospectors using gold pans, sluice boxes, and metal detectors.

Although it may seem that today the business of gold mining is reserved only for large companies, there are many thousands of individuals around the world who pan and mine for gold in different locations around the world. These artisanal miners find gold all across the globe in varied climates and ecosystems.

Prospectors can use a variety of methods to find gold deposits, but a proper understanding of how gold is naturally deposited in the environment, as well as the natural processes that cause it to accumulate in certain areas, are what will really help the modern day gold miner learn how to successfully find gold.

Basic Geology to Locate Gold Deposits

Not all gold deposits out there have been found and mined. There are still places that contain gold that have never been worked, despite the fact that miners have been searching for gold for hundreds of years now. If you can find one of these undiscovered gold deposits there is a good chance that you have found something special, since you will have been the first person to mine it.

To find one of these deposits, you need to be able to identify the natural indicators that will lead you to finding the gold. Prospecting blindly will not be productive. Even a basic understanding of geology and the indicators that can lead you to gold will put you leaps-and-bounds ahead of the average prospector.

Productive Rock Types for Your Area

When researching a mining district, good mining reports will indicate the general rock types that are associated with the productive gold mines in the area. Pay attention to these most common rock types and always be looking for them when you are out prospecting. They can be an indicator of where gold will occur.

Probably just as important to identifying the area rocks associated with gold is being able to identify the host rocks that are generally not associated with gold. If your research has never indicated that gold occurs within a certain rock type, then you certainly don't want to spend a significant amount of time searching within that type of geology.

Geological Contact Zones

Being able to identify geological contact points is very important and often completely overlooked by prospectors when searching for areas where gold will occur. This is an area where two different kinds of rock come together. The easiest way to find these areas is to study geologic maps that show the boundaries of different rock types.

Your research will often indicate which rock types are going to be the most productive. Often it's not the rock type specifically, but rather the contact point between two large rock types. That contact created pressure and extremely high temperatures which caused fissures to form and gold to be pushed up to the surface.

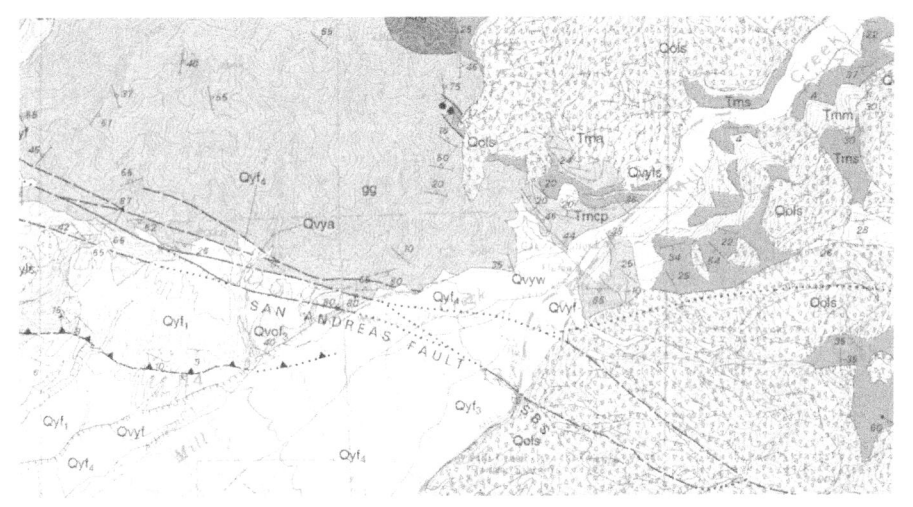

Geological maps can be extremely useful to the gold prospector. Use them to identify rock types and contact zones associated with the established mines in the area. This is the type of research that can help you locate virgin gold deposits that other miners have missed.

The general trend of the geology within your district can be very important. Look for contact points where different rock types run perpendicular to the general trend of the mountains. These contacts resulted in high pressure and high temperature conditions that commonly produce gold. You will find that many of these areas will have some historic workings, plus there are still areas out there that are "textbook" contact zones that are very rich with gold that have never been mined.

Soil color changes in the soil are another indicator of a contact zone. Depending on how much bedrock is exposed in an area, you may or may not be able to easily identify contacts where different rock types come together, but you will be able to see where soil color changes. Since soil is composed of the host rock, even a small change in soil color can be an excellent indicator of a contact zone.

Some color changes are very readily apparent, while others can be quite subtle. You aren't looking for small areas with minor change here; you want to try and identify distinct lines of different soil types. Sometimes these places can be identified using aerial imagery, but in places with ground cover like trees and shrubs, it usually takes boots on the ground.

These contact zones may be generally short, but sometimes they will run in a mostly straight line for many miles. You may also have success finding new gold bearing areas by locating productive mines and then noting a color change that extends off from the mine. There may be valuable gold deposits in a nearby drainage that are an extension of the same contact zone that occurs at a well-known mine just over the hill.

Iron Staining - Hematite - Magnetite - Black Sands

If you have done any amount of gold prospecting, you probably know that gold and iron have a very strong relationship. Gold is almost always associated with iron. When you pan for gold and find black sands among the fine gold, it is generally comprised of hematite and magnetite. These are both types of iron oxide that are common to almost all gold bearing areas. You will also find that most of the areas that you can find gold nuggets with a metal detector also have high iron content.

This is easily visible by the presence of very dark soils. They are often black or reddish in color, but they can even show purple, orange, yellow - a variety of different colors. These dark or brightly colored soils can be an indicator of high iron content, as well as many other minerals associated with gold.

This is the reason that it is recommended that you use metal detectors specifically designed for detecting gold nuggets, since they are specifically designed to handle these highly mineralized, iron-rich environments.

Quartz

Most people know about the common association of gold with quartz. Gold veins often form within quartz rock and it is certainly an indicator to look for. However, many prospectors give more attention to quartz than it really deserves.

Quartz is the second most abundant mineral on the Earth's surface, and it can be found in many locations that have very little or no gold. Thus, **the presence of quartz by itself is a fairly poor indicator of the potential for gold.**

While the presence of quartz alone is not a very good indicator of where gold can be found, there is no doubt that there are many gold districts where gold and quartz have a strong correlation.

I generally consider quartz to be a good indicator once I know that I am in a known gold bearing area, and I have identified that there is a relationship between gold and quartz in that particular area.

Let me explain that a bit more for clarification. There are some gold bearing areas where gold and quartz are commonly found together. Many times the gold nuggets that are found will have a very coarse texture and still have quartz attached to them down in the grooves of the nugget. This indicates that they were eroded directly from the quartz. In these areas, it is worthwhile to spend some time prospecting around quartz outcroppings and scanning quartz pieces with a metal detector.

However, there are many areas that you can find gold nuggets that seem to have little to no association with quartz. There may be quartz present in the area simply because it is so common, but the gold doesn't run through the quartz itself, it is just there.

The kind of quartz that gold is generally found in is not pure white. Most commonly it will show significant iron staining, and the quartz will have a dirty appearance with reddish/brown stains. Gold can be found in pure white quartz, but it is much less common.

Similar Appearances to Nearby Gold Districts

One of the best ways to find new, undiscovered gold deposits is to study the geology of known gold districts and then explore the fringes of that known district. Identify areas that have a similar geological appearance.

Nowadays, this is the best way to find truly undiscovered gold deposits. Even with an advanced knowledge of geology, a prospector in the 21st century is going to have a very difficult time finding a rich gold deposit in a completely new area. The places with the best odds are those areas that are close to the existing mining districts.

This could be one of several natural indicators that are similar to a gold district with a known history of gold production. It is one of the best ways to find an area that nobody has ever prospected before, but it can take a lot of time and patience, and you will likely spend a lot of time searching before you stumble upon any gold.

Gold vs. Pyrite - Identifying "Fool's Gold"

Gold and pyrite are completely different minerals, but because of their similar colors they are easily confused. Beginning prospectors look in their gold pan and see a lot of shiny yellow color and think they have hit the mother lode! Learning how to identify both gold and pyrite is one of the first skills that any gold prospector should learn. The structure, color, hardness and specific gravity are all indicators that will help to differential between the two.

The most obvious way to tell the difference between them is the specific gravity. When panning out material, gold will settle and concentrate in the bottom of the pan, but pyrite will move freely in the pan. You will often see them at the surface mixed in with the lighter sands and gravels. Proper gold panning will easily separate the two, as the small specks of pyrite will wash out of the pan while the denser gold particles will be retained.

The color is also a good indicator between the two. While gold obviously has a golden color, pyrite generally has a brassy and shiny coloration. It has shiny surfaces that catch the reflection of the sun. If you move your gold pan in a circular motion in the sunlight, gold will maintain a consistent color, while pyrite will flash in the sunlight and be dull when shaded.

If you take a pocket knife and separate the particles out, their different hardness will be readily apparent. Pyrite is much harder, so if you smash it with the tip of your pocket knife it will shatter into several pieces. Gold on the other hand is a very soft and malleable metal. It can be smashed flat without breaking apart.

The structure of each mineral is quite different, although this can be difficult to see if you are only dealing with small flakes at the bottom of your pan. If the pieces are large enough, you will notice that most pyrite is generally structured in cubic, octahedron and pyritohedron formations. Although there are crystalline gold specimens that sometimes share these rare formations, they would be considered the exception rather than the rule. Most gold nuggets and flakes that are found in rivers and creeks are polished and worn fairly smooth.

The differences between gold and pyrite are fairly obvious, and a little experimentation will clearly show you what you are dealing with. The easiest indicator that most experience gold prospectors use is simply evaluating the characteristics of how they react when swirled around in a gold pan. If they will easily move and remain on the surface of the material in your pan, you are likely dealing with iron pyrite. If you can see it shining in the bottom of a creek, on the top of all the sands and gravels, you have most likely got pyrite. Most experienced miners don't even give it a second thought. Once you have seen real gold, you will never confuse it with anything else.

Mining Placer Deposits

Placer vs. Lode

What is Lode Gold?

The formation of gold doesn't start in a river. It generally starts as a vein in rock. This is referred to as "lode gold" and this is the type of gold that nearly all commercial mines are after these days.

There are many different types of lode gold deposits, which all require specialized mining techniques to mine and extract. For simplicity, it's fair to simply summarize lode gold as being contained within rock.

What is Placer Gold?

Placer gold is the gold that most prospectors are more familiar with. This is the dust, flakes and nuggets that you can find by panning and sluicing a creek or river. Placer gold might have some rock attached to it, but generally it will be free of most rock material and will be worn relatively smooth. Erosion creates that typical "nugget" shape that we are all familiar with.

Placer gold accumulates over time when it erodes from hard rock veins. Gold doesn't form in the river, it is transported there and forms concentrations. The origin of that gold is usually from a vein up on the hillside.

Here is an example. Through geologic processes millions and millions of years ago, a gold vein forms in some type of rock. This vein is 10,000 feet up on a mountain and 20 feet under the ground. As time goes by natural erosion takes place. The rock over the vein is eroding away, and after years it slowly crumbles away. Now we have an exposed vein of gold on the surface of the ground. It doesn't take long for nature to take its course and a chunk of gold breaks off of the vein.

Wind and rain occur, and this course chunk of vein material slowly erodes and starts to move. A once coarse specimen of gold in quartz starts to tumble and smooth. The quartz starts to break off and heavy rain events slowly move it downhill. Eventually it makes its way into a small drainage. High water events continue to tumble and push it downstream. Millions of years go by, and a once coarse specimen becomes a smooth, polished gold nugget. What started as a gold vein eventually turns into placer gold.

Which Type is Better for Miners?

Gold is gold, so there isn't necessarily a better type of gold. However, mining for lode gold has many more challenges than mining for placer gold.

Placer gold mining is much easier and usually more productive for the average prospector. The gold is already separated from the rock, and gravity has concentrated the gold into creeks and rivers where it is more easily accessible. All the prospector needs is some simple tools, a gold pan and maybe a sluice box and they are ready to start mining.

Lode gold takes a lot more equipment and generally a lot more expense. Since the gold is locked up in rock it is usually pretty hard to get to. Even if a miner locates an exposed vein on the surface, following that vein and processing the ore as it goes into the Earth is quite challenging. It will often be much more expensive to mine these deposits than the gold that you will recover is worth.

For the average gold prospector it is usually better to spend your time focused on placer gold.

Types of Placers

The natural processes of erosion result in a variety of different ways that gold will concentrate in the natural environment. Depending on how they were formed, placer deposits are given different classifications. Below are some common placers frequently associated with gold deposition.

Alluvial Placers - these gold deposits are the most commonly found throughout the Western U.S. and were typically the first deposits that were exploited by the early gold miners. They are gold concentrated by streams and rivers, typically consisting of paystreaks on the inside bends of flowing waterways. For the most part, in the United States these deposits have been worked out on a commercial scale, with the exception of Alaska and a few other remote locations. Alluvial deposits are rejuvenated constantly to a small extent, and are some of the most popularly mined areas worked by recreational prospectors.

Eluvial Placers - these placers generally consist of deposits that form downhill of the original lode source. The forces of gravity and downhill creep move material downslope, concentrating heavier and larger concentrates toward the base of the exposure. The extent and spread of eluvial placers can vary greatly. The main lode source can commonly be located, but in some instances the entire lode deposit has weathered away, leaving behind only the eluvial deposition.

Bench Deposits - these are remnants of other ancient placers. These were typically alluvial placers at one time, but they were left "high and dry" by the down-cutting of a river system or the raising of mountains over millions of years. Rivers that are deeply entrenched in bedrock really don't move much, but many rivers that have a wider, meandering path will change location frequently with each high water event. These changes may be as small as a few feet or sometimes they might move thousands of feet. When this happens, rich placer deposits can get "left behind" by the river, and remain in place away from the current river channel.

Bench deposits are often overlooked by other prospectors. It seems that most placer miners are focused on the water's edge, and don't take the time to look around and study the surrounding area. Even in areas that have been prospected hard over the years, bench deposits are often left ignored and nearly untouched.

Residual Placers - the continuous erosional effect on a gold outcrop will result in the deposition of gold in the nearby vicinity. Lighter materials will be taken away by wind and rain, leaving the heavy mineral concentration. Although these placers are generally not extensive enough to attract commercial mining endeavors, they can be very productive for the individual prospector using a metal detector to find gold.

Beach Placers - fine gold deposits can be found along the beach sands in many locations throughout the world. Two well-known areas in the United States include the rich deposits of Nome, Alaska and the beaches of Southern Oregon. Gold is either carried to the ocean by rivers and creeks from nearby sources, or eroded directly from wave action along the beaches. These deposits can often be found directly along the shoreline and along ancient shores well above current sea level. These deposits generally contain fine gold.

Glacial Deposits - these placer deposits are formed by glacial movement, transporting gold bearing gravel from different sources and depositing them elsewhere. Glacial gold deposits are very well known throughout the Midwest and Northeastern U.S. They are almost always characteristic of very fine gold that has been pulverized by glacial action. Although widespread, these deposits are generally small and not economically viable for commercial mining endeavors.

Eolian Placers - typically found in arid regions, natural erosive processes (wind) cause sand and other light materials to blow away, exposing heavier minerals, essentially exposing the vein by eroding lighter material from around them. Although these can be rich sources, the spotty distribution of gold is generally not workable on a commercial scale. Patch hunters using metal detectors can often work eolian placers very efficiently.

Flood Gold - Very fine gold can be transported considerable distances by high water flows. Although some small amounts of concentration can occur, flood gold is almost without exception extremely fine textured, and do not concentrate in any payable quantities. Even to a recreational prospector, flood gold is very difficult to retain and does not attract much attention due to its small size and limited abundance.

Ancient River Channels - Similar to bench deposits, the presence of ancient river channels is something that many prospectors will overlook. The exciting thing about locating ancient rivers is that they are sometime virtually unknown and may have never been searched by other prospectors.

So what is an ancient river? Imagine for a moment that we are in the Jurassic time period, approximately 200 million years ago. A rich gold bearing river is flowing, a river that has never been prospected and has been completely untouched by man. As the years go by, the Earth changes in dramatic ways. Tectonic plates come together and create the mountains that now exist across the western United States. Some areas that were once valley bottoms are now located high up on the side of a mountain.

Many people have a hard time grasping the concept of geologic time and just how much the Earth can change over millions of years, but it most certainly has. There are ancient rivers that were once full of water and rich with gold that are now found high above the existing water line. Of course they no longer contain water, but the ancient river channel (and gold) remains.

GOLD DEPOSITS

Inside Bends of streams and rivers help to concentrate gold in low pressure area.

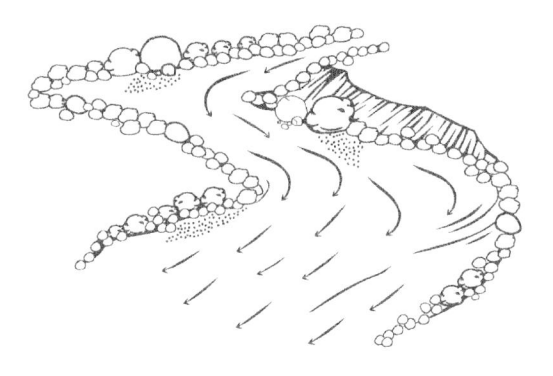

Gold deposits will form behind obstructions such as large rocks and boulders.

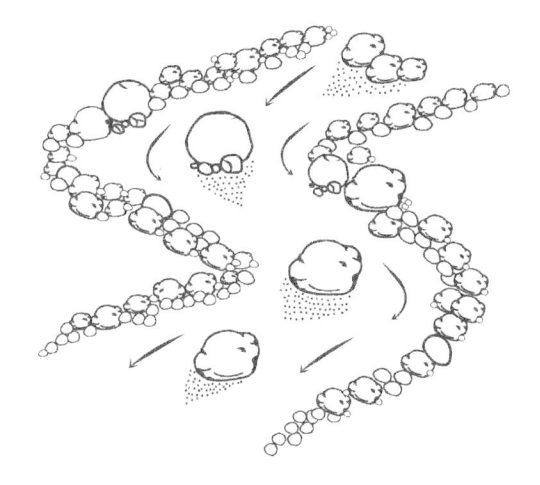

How Gold is Deposited in a Waterway

Gold has a higher specific gravity than the sands and gravels in the stream, and that makes it somewhat predictable. You need to go down to find it, down below the sands and gravels, and **down deep into the cracks and crevices in bedrock**.

During a high water event, the material in a stream acts in the same way that it does in your gold pan. Everything is being agitated, and the heavier gold is able to travel downward until it hits a major obstruction.

Beginning prospectors commonly make the mistake of not digging deep enough to get to the gold. It may be right below their feet, but it is often covered by significant overburden that contains little or no gold. Not only does gold go deep down in the gravel, **the gold is also constantly moving downstream until it reaches an obstruction**. It may get caught in a deep bedrock crack, but before it settles down to bedrock it has a good chance of moving a significant distance down the stream during a high velocity stream flow.

Once again, it's weight will cause it to settle in certain predictable areas, places where the stream slows down or loses the energy needed to move the gold any further. These areas include deep pools beneath waterfalls, behind large rocks and boulders, among exposed tree roots, and the inside bends of a stream. If you are a fisherman, then a good way to spot areas like this is to identify places that fish would naturally congregated. The slack water near the faster current. These are the same general areas where gold will lose momentum and get hung up in bedrock.

When prospecting for gold, always be on the lookout for black sand concentrations. Black sands are composed of various iron oxides, and are commonly associated with placer gold due to their high specific gravity. These black sands will settle in a stream the same way that the gold does.

MINING PLACER DEPOSITS

IDEALIZED CROSS SECTION SHOWING FORMATION OF COMMON PLACER GOLD DEPOSITS.

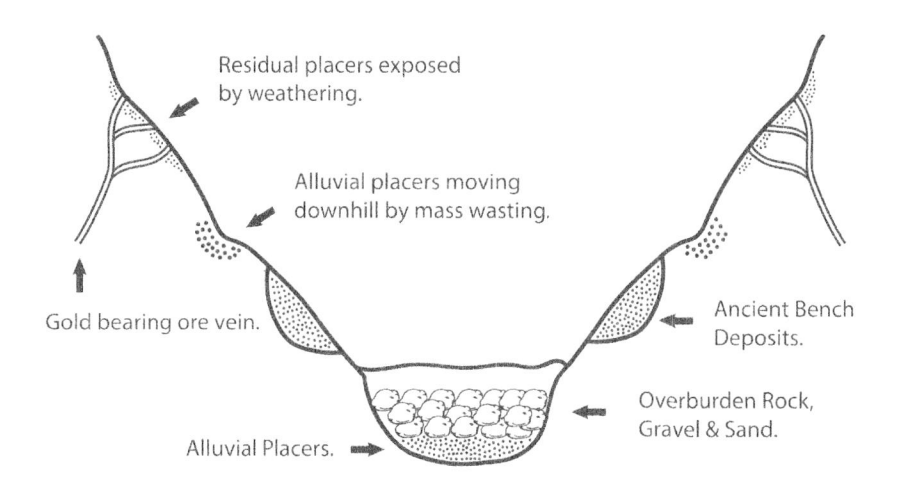

Residual placers exposed by weathering.

Alluvial placers moving downhill by mass wasting.

Gold bearing ore vein.

Ancient Bench Deposits.

Overburden Rock, Gravel & Sand.

Alluvial Placers.

IDEALIZED CROSS SECTION OF FLOWING STREAM.

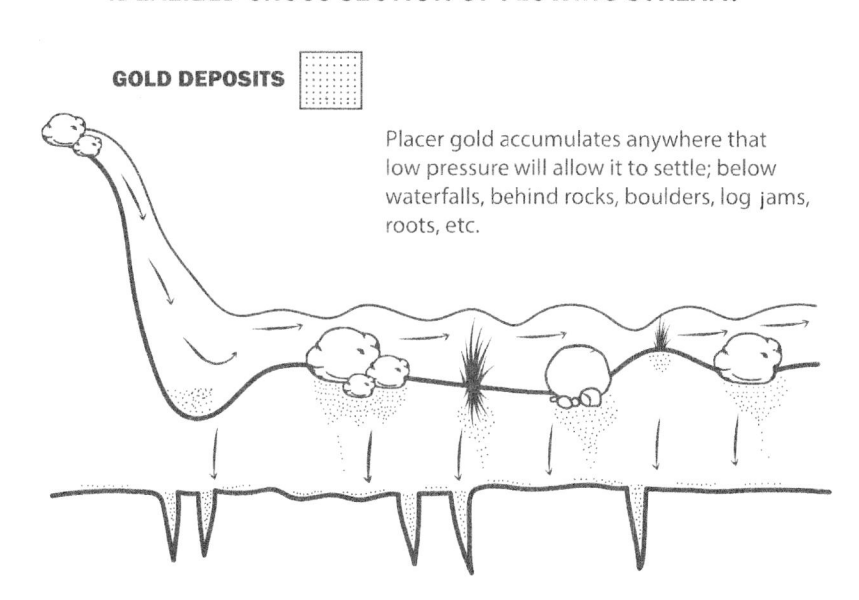

GOLD DEPOSITS

Placer gold accumulates anywhere that low pressure will allow it to settle; below waterfalls, behind rocks, boulders, log jams, roots, etc.

If you are finding gold then you are probably also finding black sand. But if you are finding black sand it does not necessarily mean that you will find gold. **The presence of black sands does not guarantee the presence of gold, but if you are in a gold bearing stream there is a good chance that gold is nearby.** If your gold panning is resulting in black sands, mercury, and lead bullets or fishing weight, you are on the right track.

Placer gold deposits are usually not far off from where lode gold deposits are. The gold in hard rock is released during the weathering process and then it is carried off by glaciers or running water (streams, rivers, flood water) to someplace where it is deposited and concentrated in a placer deposit.

Because gold is relatively heavy, it is generally not carried far from the source. This means that prospecting rivers and waterways around areas where lode deposit has been found may result in the discovery of gold. In addition, ancient river channels in places with gold-bearing rocks and flooding outwash in such places might also contain placer gold.

The most common places to find placer gold is on the river beds. Usually when a gold-bearing vein getst exposed due to the weathering of the top rocks, and other forces of nature, the gold in the vein doesn't move far from the source. However, when a large storm hits the area causing rivers and streams to flood, the runoff water runs faster, carrying with it loose gold particles and nuggets into the streams and rivers.

Sometimes a river or stream flowing over a gold vein may eventually cut deep enough to erode the vein and carry off loose gold particles and nuggets downstream. Gold, being heavier than all the other materials being carried downstream, exerts a downward force on the water and quickly settles on the riverbed once the water slows down. This creates a rich placer gold deposit on the river bed.

Where Can I Legally Prospect

Land Ownership

Once a general gold-bearing area is identified, finding a specific area to start prospecting can sometimes be a serious challenge. The two most common barriers that prospectors will encounter are private lands and mining claims. In some mining districts, finding an area that is open to legally prospect can be quite challenging.

It is each prospector's responsibility to make sure they are in an area where they are allowed to prospect.

There are a few simple options that new prospectors can take to gain access to gold bearing ground. The first thing that I would recommend is joining a gold mining club. The best known mining club is the Gold Prospectors Association of America. The GPAA has chapters in almost every state, and the benefit of joining is that you will have access to mining claims that they hold throughout the U.S. Most states have at least a few other local clubs as well, most of which have a handful of claims that members can prospect on. Joining a club allows any prospector to be able to work private claims for a very reasonable cost, and is highly recommended for beginners.

Another option that surprisingly few people seem to consider is simply finding the owners (of either private land or mining claims) and asking their permission to prospect. Often private landowners own land that has gold, but the landowners themselves are not miners and have no interest in gold prospecting. Simply ask permission to prospect on their property and you may get a simple "yes". You can also come to an agreement to share a percentage of profits from the gold recovered with the landowner.

Asking permission from a claim holder is also a simple option that may pay off. Of course not all claim holders are going to allow others to prospect on their claims, but you don't know until you ask. I have found it is quite common to get permission if you make it clear that you are just using simple tools such as gold pans and sluice boxes. Realistically, a person using small scale mining equipment are not going to have a serious impact, and many claim holders will have no problem giving you permission to do a bit of "pick and pan" prospecting for the weekend. You may also come up with an agreement to share a percentage of the finds, as discussed in the previous paragraph.

Finally, it should be emphasized that not ALL of the gold bearing ground is claimed. Certainly in some of the more populated and well-known mining districts it can be hard to find areas that are not claimed, but I assure you that it is still possible to find areas on public lands that are unclaimed and open to prospecting. Doing the proper research will help you find these areas. Yes, most of the well-known gold areas are claimed, but don't ever believe the statements that all of the good gold ground is claimed up. Plenty of gold can still be recovered in relatively unknown areas, often on the fringes of the better known gold districts.

Private Lands

Private lands can potentially offer great opportunities for the casual prospector, and some of the best gold-bearing areas left today are currently located on private property. Of course, you can't just go on to someones property and start digging for gold. You need to get permission from the landowner.

This is an added challenge, but also an opportunity for those who are willing to put in the extra work.

BLM and Forest Service (Public Land)

For most of us (particularly in the Western US), public lands offer the best opportunities to go out and look for gold and other valuable minerals. There are literally hundreds of thousands of square miles of public land that is available to prospecting. Most of that ground spans a dozen western states and Alaska.

National Forest lands provide millions of acres of public land for gold prospecting.

Most opportunities for the small-scale gold prospector are going to be found on lands that are managed by the US Forest Service and the Bureau of Land Management. These lands are open for all types of outdoor recreation, and they offer countless opportunities to go out and explore, prospect, and potentially discover rich mineral deposits.

There is an added layer of complexity regarding public land prospecting, and that is the existence of mining claims. You will need to be aware of existing claims to avoid illegally trespassing on another miner's claim. However, claims are also a fantastic opportunity as you also have the ability to file your own claim if you discover a mineral deposit.

National Parks

National Parks are generally off-limits to collecting almost everything. Panning or sluicing for gold will likely result in a large fine. The only exceptions I am aware of are some of the National Parks in Alaska that do allow *casual use*, allowing you to pan for gold and collect small rocks. However, no hand tools are allowed and certainly no large equipment.

Wilderness Areas

Wilderness areas are also generally off-limits to gold prospecting and rockhounding. There are exceptions for some historic claims that existed prior to the wilderness designation, but with few exceptions it is best to assume that designated wilderness areas are not open to any type of mineral extraction.

Tribal Lands

As a general rule, you will want to avoid tribal lands. In my experience, unless you are a tribal member you will not be able to get permission to prospect. I suppose it can't hurt to contact them and ask, but don't get your hopes up. I've never met anyone who was ever able to get permission, and I know many who have tried. It's too bad, because there is a lot of rich historic mining ground that is now located inside of Indian Reservations.

Mining Claims

Determining Claim Status

Mining claims are probably one of the most misunderstood (and important!) aspects of gold prospecting. In the United States, a mining claim gives the claim holder exclusive rights to the minerals on a predetermined tract of land, assuming they have properly staked, claimed, and maintained the claim by paying their required fees.

The terms "claim jumper" and "high-grader" aren't just from days past. There are still thousands of acres throughout the United States that are claimed, and you can get into serious trouble if you go prospecting or metal detecting on someone's claim without their permission. Despite popular belief, a mining claim does not need to be marked and posted to be a valid claim. Claim markers often fall down over time, get run over, or stolen. It is very common to have claims posts still up in places where a claim has expired and is now open for mineral exploration again. The previous claim holder just didn't bother to go out and remove their claim markers.

You need to determine if you can legally search for gold in an area prior to ever leaving the house. One of the biggest challenges with gold prospecting is simply finding an area that is open where you can legally prospect. With that said, there are plenty of good gold-bearing areas where you can find gold, and you just need to go to the extra effort of finding them.

A program called the LR2000 is the best online tool available to search for currently claimed status (links below). Claims are filed with the state BLM office as well as the County Recording Office. It is always best to contact these offices for further information and clarification about mining in your area.

LR2000 info
https://reports.blm.gov/reports/LR2000

Staking Your Own Claim in 8 Steps

1) Select an Area with the Minerals you are Interested in

Before you even think of staking a claim on the federal lands you first have to find out if the land has the particular mineral of your interest or not. Most commonly this is silver and gold, but prospectors can also stake clams for copper, platinum, oil, asbestos, lithium and all sorts of metallic and non-metallic minerals.

Once you've found an area that potentially has gold, check the land status to make sure that the areas is federal land that is open to exploration. There's no point going any further than this if you find out that the land is restricted and off-limits to mining.

2) Check if there is Another Active Claim on the Land

Once you have found an area on the federal lands with minerals worth staking a claim, it is important that you confirm with the local BLM offices that there is no active claim on the land before you stake your own claim. If there is a claim on the land it should be marked in accordance with the specific state laws.

Some claims may not be well marked but may still be valid because claim posts can be lost, stolen, or fall down and become hidden. Make sure also to check if someone else is in the process of staking a claim on the same location as your desired area. It is important to check on this to avoid conflict that may arise as a result of your claim staking activities.

3) Stake Your Claim

Once you've found an open section of federal land, prospected the area, and located a potentially profitable mineral deposit, it's time to actually stake your claim.

You start to stake your claim by physically marking the boundaries as required by the laws in the state where it is located. Generally this is done by piling some rocks to make a marker or using a post. Different states have different rules regarding how to properly stake and how the boundaries need to be marked.

4) Plot the Location of your Claim on a Map

Now that you have located and staked your mining claim on the ground, the next step is to identify your mining claim location on a map for filing purposes. You should have already done this when you were researching open areas to prospect, so this part should be easy. The map of your mining claim should show the range, the meridian, the township, the section and the land ownership. It should be well detailed for easy identification by anyone else looking for mining claims in the same area.

5) Fill out the Government Paperwork

Before the claim becomes legally yours, you need to fill out some paper work with both the federal and the local county where the claim is located. State laws in most of the states require that you first file the original claim location notice in the county clerk's office of the county in which the claim is located. This is very important! If you just deal with the BLM and don't also record your claim with the county then your claim will not be valid.

Different states have slightly different requirement for filling of the location notice. Usually, the filling has to be done within 90 days after staking a claim on the ground. Take note that some states require early filing possibly within 30 or 60 days, which means that you should do it soon enough to prevent your claim from being taken over by someone else.

Once you have filed the location notice with your state government you can now go ahead and file the same with the federal government through the BLM offices. This should be done in accordance with the Federal Land Policy and Management Act (FLPMA) of 1976, which requires that you file a copy of the official certificate of location or an official record of the notice with BLM.

6) Surface Use of your Claim

Once you have successfully staked a mining claim, you probably are excited to start mining on it. Wait! Before you start any mining activity on a new mining claim you should check with the local authorities about relevant regulations regarding mining activities in the area. Check the state laws on mining and the BLM mining regulations on surface management mining, and the Forest Service. Just because you have a valid mining claim does not mean you can do any type of mining you want. There are likely additional requirements if you want to use heavy equipment, build ponds, roads, or use suction dredges.

7. Amendments and Transfers of a Mining Claim

A mining claim anywhere in the United States can be transferred either in part or in its entirety. The first notice of the transfer should be done by the state government which will issue the required documents such as the quitclaim deed for the transfer.

Other than a transfer, you can also make an amendment to the description of your claim location. Transfer and amendment should be done at both the county office and the state BLM offices. The transfer should be done within 60 days after the transfer of the claim; failure to to do so will render the transfer null and void. You must be prepared to pay a service fee for this notification.

8. Abandonment or Relinquishment

Once you have successfully staked a mining claim you can use it for many years provided that you pay the yearly renewal fee on time. Failure to pay the fee on time will mean that you have relinquished your rights to the claim, so make sure that you keep on top of this. Many miners have lost access to their mining claims by carelessly forgetting to file their annual paperwork.

Constantly Changing Mining Regulations

The Modern Miner

Most of the state agencies seem to be working in the direction of making gold mining a "hobby" activity. But since the beginning of time, mining was a job. Men didn't travel around the world to find gold just because they thought it was pretty. The were extracting wealth out of the ground and trying to make a living.

Finding paying quantities of gold was certainly easier during the Gold Rush. Placer deposits were completely untouched, and as a result a man could easily find ounces per day in the right location. This could often be done with the simplest of tools like gold pans and sluice boxes. But the thing that made mining truly viable during the early gold rushes was the use of large-scale mining techniques. You could divert water, build small reservoirs, hydraulic mine, bucket dredge, use explosives and do just about anything else that you wanted if it would help you extract the gold. Most of these tactics are now illegal, or they are so heavily regulated that it's next to impossible for the average person to do it today.

Now I'm not necessarily saying I wish this stuff was allowed today. There were serious environmental impacts to a lot of these old methods. However, we have taken such a drastic 180 degree turn in the past century that it's now extremely difficult for the average prospector to make a living wage from gold mining.

Motorized Mining Equipment

Suction dredging is a prime example of what has become so heavily regulated just in the past few years. For many decades, the suction dredge was the single most efficient way to mine an alluvial gold deposit. Their design allowed a miner to don a wetsuit and suck gravel right from the bottom of the river. The material would run through a sluice box floating at the surface, and the gravel would kick out the back and be returned to the waterway.

No material is removed (except gold and small amounts of "heavies") and nothing is added. The environmental impact is extremely small. And "damage" caused by dredging a hole down to bedrock is absolutely trivial when compared to the amount of gravel that naturally moves every year during spring flooding.

State-by-State Restrictions

The states vary a great deal, but unfortunately there have been major shifts in anti-mining efforts in just the past few years. California has been a problem for decades, but now even states like Idaho, Oregon and Washington are passing new rules and restrictions that make it hard for a person to get out and dig. Many waters are particularly challenging because of the presence of endangered salmon and steelhead, which are used as a reason to add excessive restrictions.

I won't go over the rules of each state because, quite frankly, they are almost certain to be different (and worse) in a very short time. Some states like Alaska, Colorado and Nevada are still relatively "miner friendly."

The trend is to move to smaller and smaller equipment to mine for gold, using our bare hands mucking around in the dirt like a child. Yes, the dreaded "hands and pans" areas seem to be expanding more and more all the time. It's important to keep up on updates to mining regulations in your state. The fines can be severe.

Gold Prospecting Equipment & Use

To increase your odds of success, I recommend putting together a good basic prospecting kit, a collection of tools that will allow you to recover as much good quality material as possible. These tools will come in handy regardless of your prospecting method.

Basic Prospecting Kit

Small Crevice Tools - The most important pieces of your kit should be a variety of smaller tools that can be maneuvered into small cracks. Lots of simple everyday items work well for this. A kitchen spoon, heavy knife, screwdrivers of various sizes, and commercially produced crevicing tools are very handy.

Pry Bar - Bring along a big pry bar or crowbar to help open up those deep, narrow bedrock cracks. This is the key to reaching gold that has been missed by others. You would be amazed at the amount of placer gold that is still hidden in rather obvious locations where no one has taken the time to work them properly.

Standard Shovel - You'll need a basic shovel to remove overburden and get down to the richest material. Get a good quality one so it doesn't break.

5-Gallon Bucket - A few buckets are very helpful for hauling dirt.

Snuffer Bottle - Snuffer bottles help you capture the small particles of gold at the bottom of your gold pan. Use one to gently blow away lighter material and expose tiny bits of gold.

Hand Trowel - Once a crack is opened up a bit, a small hand trowel can be very handy. Again, smaller is usually better. I have taken an existing garden trowel and cut the sides off, so that it is only about 1" wide. It's one of my favorite tools.

Hammer and chisel - These are used to work open cracks. Tiny cracks in bedrock often hold the best gold, and you need to open them up to clean them out effectively.

Tweezers - You'd be surprised how often you will be doing some crevicing and spot a nice nugget right out in the open where you are working. A pair of tweezers will do the trick to pick that little guy out of the crack.

Gold Pan - Of course you'll need some way to process all that gravel after you dig it up. Even if you are using larger equipment, *every prospector should have a gold pan.* The pan is a necessity for sampling and clean-up. Even multi million dollar placer mining operations have gold pans on hand.

Gold Bottle - You need some way to keep all that gold you find. Any small bottle will work, but I recommend something durable. I use waterproof match containers. They are hard plastic and bright orange. You can find them at Army/Navy stores.

Tweezers   Snuffer bottle   Gold bottle

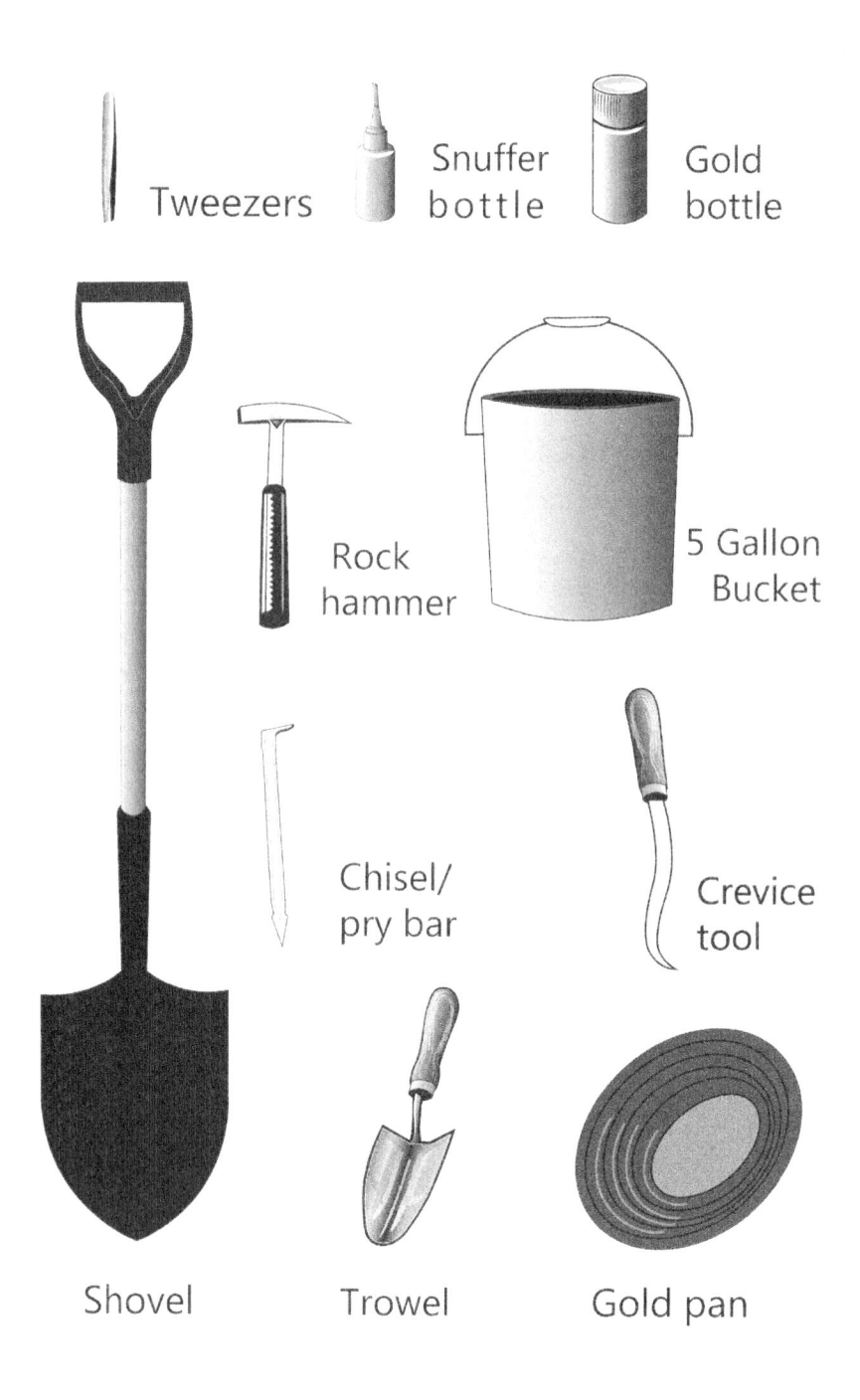

Rock hammer

5 Gallon Bucket

Chisel/ pry bar

Crevice tool

Shovel   Trowel   Gold pan

Selecting a Gold Pan

Choosing a gold pan is something that many prospectors don't really give a lot of thought to. This is most likely because gold pans are inexpensive items, and at first glance they all seem more or less the same. However, if you take some extra time to choose the proper gold pan you will significantly improve not only your success in finding gold, but your overall enjoyment of the hobby as well. So let's take a quick look at a few of the things you will want to consider before choosing a pan.

The most important thing to consider when selecting a gold pan is its size. Gold pans come in many different sizes, with diameters most commonly ranging from 10" to 18". Many beginning prospectors make the mistake of picking the largest gold pan that they can get their hands on. After all, we all know that the more material that you can process, the more gold you are likely to recover. The problem with these large pans is that they are just too much for the average prospector to use comfortably. When an 18" pan is loaded up with gravel, it gets very heavy and tough to maneuver, turning the pleasurable experience of gold panning into a real chore. Now if you are a large man in good shape, maybe the larger sized gold pans are ideal for you. However, children, women, and most men will probably find that a small to medium size is a better choice.

To figure out which size is best for you, lay the gold pan face down on the inside of your forearm. With one edge of the pan at your elbow crease, look at where the other side of the pan is. Does it extend past your fingers? If so, then it is definitely larger than you want. The ideal pan will fit from the crease in your elbow to somewhere around your knuckles. Remember that a gold pan should be used more as a sampling tool than for large-scale prospecting anyway, so save your back and go with a smaller size that will make panning a pleasurable experience.

Another choice you have is whether you want a plastic or metal gold pan. Of course, we all know that the metal gold pan was the sampling tool of choice used by the old time prospectors many years ago, and certainly there has been a whole lot of placer gold pulled out of the creeks and rivers with these pans. With that said, today's modern plastic gold pans have many advantages that should be considered before making your purchase. One of the biggest advantages is maintenance, or lack of. Metal gold pans will rust if put away wet. I know that I am often tired after a long day of prospecting and forget to do something simple like drying off a metal gold pan. A few days or weeks later and you have a rusty mess on your hands.

With a plastic gold pan, there is no maintenance involved. You can toss it in your backpack and forget about it for a few weeks without having any rust issues. Plastic gold pans are also ready to use straight out of the box, while a metal gold pan requires some preparation to remove factory oils. Plastic gold pans also have built in riffles, which will really help you retain some of the smaller flakes of gold, an added bonus for a beginner who is still learning the basics of gold panning. Plastic pans will also come in various colors that can help you spot those little specks of gold in the bottom of the pan.

You will find that there are many different pans on the market. Some have slightly different shapes, they will come in a variety of colors, and most manufacturers will claim that their model is the best for retaining gold. More important than anything is selecting a gold pan that you are comfortable using, and panning in a spot that has potential to have some gold. Once you've done that, all that is required is a little hard work and you will have a good chance of finding some gold.

How to Pan for Gold

Every prospector should know how to use a gold pan. They are still the best tool for sampling, and are inexpensive to buy. Once a nice gold deposit is found, you will probably want to use some larger equipment, but for moving around and exploring new areas a gold pan is still the tool of choice.

Find an area along a stream or river that has good potential for having placer gold. The best areas to start looking are places that have a known history of gold production in the past. Use a shovel to dig deep into the gravels, continue digging until you reach bedrock, as this is where most placer gold will accumulate. Put a moderate quantity of gravels into your gold pan. Don't load your pan with so much material that it is hard to work with.

Remove some of the larger rocks from your pan, or consider using a classifier to keep the larger material out of your pan. Rinse the rocks off in your pan so any gold that may be clinging to them will go back into your gold pan. Slightly agitating the mixture with water helps to expose the largest rocks for easy removal.

Submerge the gold pan in water and agitate the gravel using a circular or side to side motion. At first, try not to let any gravel fall out of the pan. Use your hands to break up any clumps of gravel or organic matter than might be holding gold. Continue shaking the gold pan, which will allow the gold to work its way to the bottom of the pan.

Once the gravels have been thoroughly agitated, you want to begin to remove some of the material from the gold pan. Keep moving the pan in a circular motion, and slightly tip the farthest edge of the pan away from your body. Ensure that there is always plenty of water in the pan to keep the contents suspended, allowing the gold to remain safely at the bottom. Some of the lighter sands and gravel will pour out of the pan. Every so often, level the pan out and agitate the mixture to ensure that all the fine gold remains safely in the pan. Repeat this process, and eventually you will reduce the contents of your pan to just the heaviest materials.

One you have gotten down to just the heaviest contents, the separation process will get a little bit tougher. Take it slow and continue reducing the material until you are down to a very small amount of material. At this point you should be down to mainly black sands and hopefully a little bit of gold! With plenty of water in your pan, slightly shake it and try to get the black sands to separate from the gold. This can be easier said than done, but keep practicing and your skills will improve.

If you are fortunate to have any sizable flakes or small gold nuggets in your pan, use your fingers or tweezers to carefully remove them. Place them safely in a small vial for safe keeping. Use a snuffer bottle to carefully separate the black sands from the fine gold left in the pan. Then suck up the gold into the snuffer bottle. This can take a bit of practice to master, but be patient with it and you will improve your skills.

Keep in mind that not every pan is going to have gold; sometimes you have to do a little searching to find it. Keep on prospecting, look in all the likely areas, and sooner or later you will start finding some "color". Once you have found an area that is producing a good quantity of placer gold by panning, you may want to start using a sluice box to increase your production.

GOLD PROSPECTING EQUIPMENT & USE

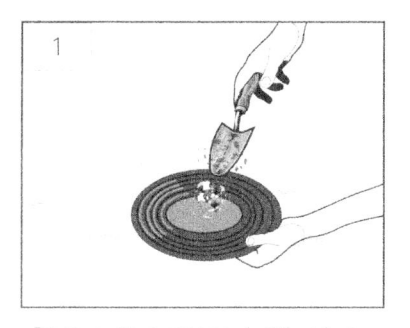

1. Dig down with your trowel and get the richest material you can. Fill your gold pan with a modest amount of gravel.

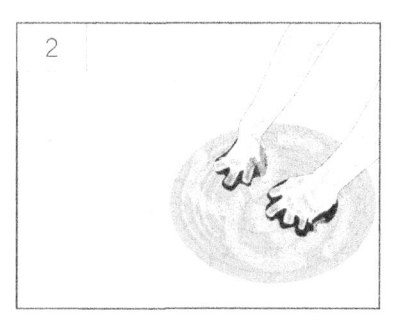

2. Submerge the gold pan underwater and agitate the gravel. Use your hands to break apart clay and remove larger rocks.

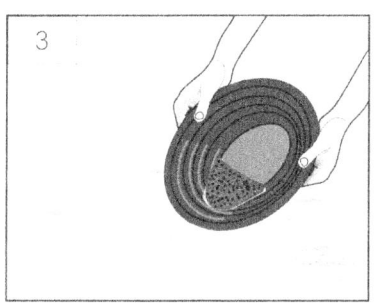

3. Begin moving the pan with a circular or side-to-side action, and gently tip the pan away from your body. Allow the lighter sand and gravel to run across the riffles and fall out of the pan.

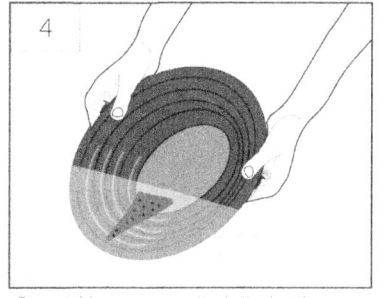

4. Repeat this process, routinely tipping the pan back upright and aggressively agitating the contents to retain the gold. Continue until the contents have been reduced down to a small amount of fines.

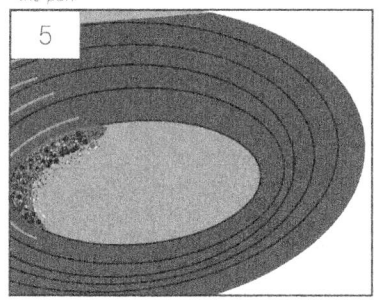

5. Once the contents of your pan are reduced to a few thimblefuls of fines, tip your pan up and swirl GENTLY in a circular motion. Take it slow. When done properly, the black sands will wash away and expose the small bits of gold.

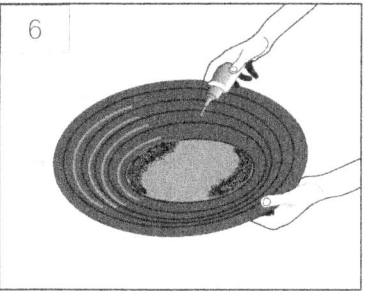

6. If you see any larger nuggets or pickers use tweezers to pick them up and place directly into your bottle. For the fine gold dust, use your snuffer bottle to wash aside the black sands and suck up those tiny specks of gold.

99

Sniping

One of the simplest and most effective methods of finding gold is sniping, and it can be a really enjoyable way to add some gold to your collection.

If you truly enjoy the process of reading the lay of the land and water to evaluate the best places to hunt for gold, then sniping is one of those prospecting methods that will really appeal to you. Basically, sniping is throwing on a mask and snorkel and visually searching for high-grade pockets where gold may be hiding in a stream. A good sniper must evaluate the area that they are working in, thinking about all the areas that gold might be hiding, and then recover that gold using a few simple tools. No heavy equipment or motorized gear needed, just some skill and knowledge about gold, and how it acts in its natural environment.

In addition to a mask and snorkel, you need to have a few basic tools to help out. Basically what you are looking for is areas that have the potential to trap and retain gold. These are the cracks and crevices in bedrock that gold can work its way into and sit for millions of years untouched. These cracks will be all different sizes, so the more varied your tools, the better chance you have of being able to reach the gold. Essentially you want to be able to clean the crevice out completely, leaving nothing when you are done.

To have any chance of doing this, bring along your prospecting kit with a chisel and hammer, a few screwdrivers of various sizes, a small garden trowel, crevicing tools, a couple of spoons, and any other small metal tools you think will work. Since these are all small tools, it's easy to have a nice assortment without too much effort. You probably have everything you need in your garage right now. You should also bring along a pair of tweezers, magnet, a snuffer bottle and gold pan.

The main idea here is to "think small". Unlike most other mining methods, we are not trying to process the biggest amount of material that we can. We are looking for the very best material and high-grading, collecting just the very richest material in the stream. Locate a promising crevice in the bedrock. Try to think of areas that have not been searched before, maybe an area that would be too difficult to access with a suction dredge or highbanker. The beauty of sniping is that with so little equipment required, it is simple to put everything you need in a small day pack and hike a mile or two to get away from the crowds.

Once you have found a nice crack in the bedrock that holds potential, it is simply a matter of cleaning it out. Of course, this is usually easier said than done. You rarely have much room to work, but it is very important to do a thorough job cleaning out the crack, because the very richest material will be found at the bottom. This is where you will be glad that you have a good variety of tools at your disposal, because no two cracks are ever the same. Use all the tools that you need, and take your time to really do it right. Often times as you are underwater cleaning out a crevice, you will get the pleasure and thrill of seeing a nugget uncovered. There is nothing quite like taking a pair of tweezers and plucking a nugget from its resting place. Clean out all the high grade material from those cracks and pan it out. It doesn't get much more basic than that, but you will be amazed at the results if you take the time to do it right.

One of the most pleasurable things about sniping is its simplicity. You can hike in and get away from the crowds, there are no motors humming or metal detectors making noise in your ear. I think many prospectors hear about this method and doubt its effectiveness. It is human nature to think that bigger is better, and certainly that is often the case when it comes to gold mining. However, there is no doubt that sniping can be a great and effective way to recover more gold. It is a fantastic way to spend a hot summer day, and if you have patience and really learn to read the stream and what it is telling you, you will find gold using this method.

Sluicing

While the gold pan should certainly be the foundation of your prospecting gear, once you start finding some gold with your gold pan, you should definitely consider using some larger equipment so that you can process more material. More material generally means more gold found at the end of the day.

A sluice box is generally the best choice. They are still relatively inexpensive; yet allow a prospector to process significantly more material than just using a gold pan alone.

How to Use a Sluice Box

A sluice box is designed to be used in a stream or river and has been used all around the world. There have certainly been improvements since the early days, but their overall concept remains unchanged. They are an excellent tool for the modern day gold prospector who is looking for a way to process more material.

There are a lot of sluice boxes on the market, and they will all find gold. The old timers didn't buy their sluice boxes, but actually made them themselves. It is a relatively simple project to build a sluice box that will find gold, but purchasing a modern design made of quality lightweight material will be more than worth the money because of the added benefit of portability. Just be sure to invest in a quality piece of equipment that can handle years of abuse, because they will definitely get worked hard. Cheap, poor quality equipment is never a good idea when gold mining.

Once you have selected a sluice box, you need to identify an area that has good flow of water and the potential to have some gold.

Setting up your sluice box will take a little trial and error if you have never done it before. What you are trying to do is find an area that has good flow that you can run through the box and over the riffles. By laying the sluice parallel to the flow of the water, you can pick various spots to set it in the stream to regulate the amount of water running through it. The ideal location will be fairly shallow with a good volume of water, which you can regulate by using big rocks to funnel the water over the riffles.

Anchor the sluice using a few large rocks, and keep the bottom end of the box slightly lower than the upper end. What you are trying to do is get the perfect amount of flow running through it, which will allow the heavier material to get caught in the riffles and carpet of the sluice, while at the same time allow the lighter materials to be discharged out of the back. If you have never done this before, it may be difficult to set up just right, and you will find that your riffles are getting clogged with sediment. If this is happening, increase the speed of the water through the box, either by slightly raising the head of the box, or by adding a few rocks to the head end of the entrance to funnel a little more water through it. With a bit of experimentation, you will get it figured out. Test it out by adding a small amount of gravel and watching how it works through your box. If it appears that the lighter stuff is getting kicked out the back and the heavier concentrates are being retained, it's time to start adding material.

At this point some gold prospectors shovel gravel directly into the head of the sluice. While this does work, your will get better results if you first classify the material that you run. By running the gravel through a coarse screen, you will remove the large rocks and debris. This larger material probably does not have any gold in it, so there is no reason to run it through the sluice. The second and more important benefit of classifying material is that it will allow your sluice to run more efficiently. Big rocks and gravel rolling through the box and over the riffles may dislodge some gold that had settled into your riffles.

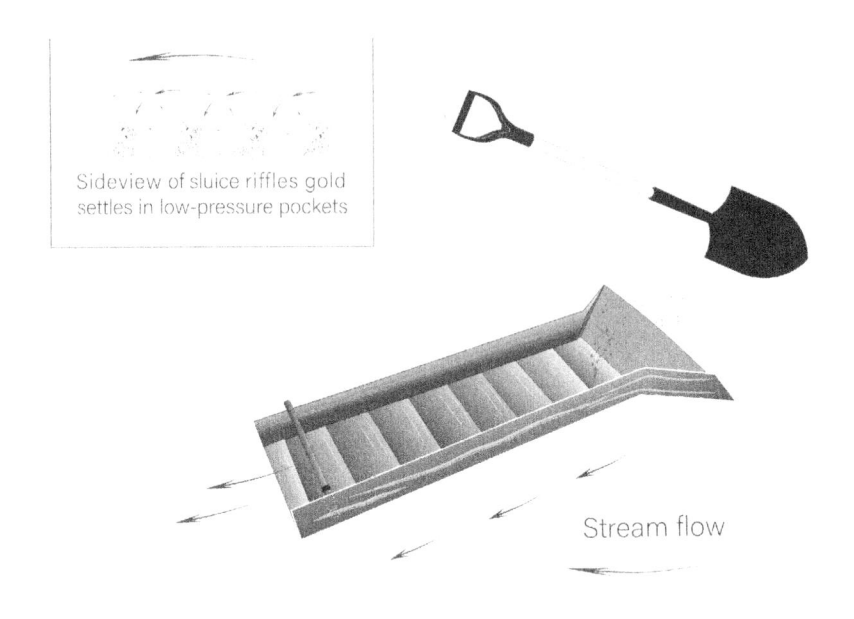

Sideview of sluice riffles gold settles in low-pressure pockets

Stream flow

When you have a few buckets of classified material, add it slowly to the front of the sluice box. Doing this slowly is important, because adding too much material too quickly will overload the riffles, preventing your sluice from working effectively. If too much gravel is added too quickly, gold can go right through the box without being retained. Using a small garden trowel for this process may help you to slow down this process. Add a little bit of material and watch it work its way through the sluice. If everything still looks like it is working properly, then keep adding material.

At some point, you will want to extract the concentrates from your sluice. Remove the rocks that you have anchoring it in place and carefully remove the sluice from the water. It is important to take care with this step, as all your hard work is retained in those riffles. While keeping the sluice box level, remove it from the water and place the bottom end into your cleanup bucket.

Check the first few riffles for visible gold nuggets and pickers. If there are any small nuggets that are big enough to pick up, carefully remove them using a pair of tweezers and place it into a vial. Any coarser pieces of gold will most likely be in the first few inches of riffles. Take apart the sluice and remove the carpeting, being careful not to lose any of the concentrates. Wash the carpet out in a bucket until it has been thoroughly cleaned up, and rinse down the screen and any other parts of the box that may hold small pieces of gold. Once everything is cleaned up, you can reassemble your sluice and put it back in the water.

It is your choice to process the concentrates now, or to take them home for the final processing and removal of gold. Many gold prospectors choose to take their concentrates home, so they can spend their valuable time at the river digging as much gravel as they can. Then when they get home, they can do the final gold recovery using a gold pan.

It is as simple as that! Sluice boxes are excellent toosl that every gold prospector should have in their arsenal. They have proven themselves as an efficient and effective gold recovery tool, and the modern designs are light weight and work better than ever before.

Highbankers, Dredges, Trommels & Drywashers

Suction Dredging

There have been many advances in modern gold mining methods over the past 150 years, and today the suction dredge is one of the most efficient and effective gold recovery tools used by the modern day gold prospector. A suction dredge essentially works like an underwater vacuum cleaner. It sucks up the stream bed materials using a suction hose and processes it through a recovery system floating on the surface of the water. The recovery system separates and retains the heavier materials (gold, black sands. etc.) and discharges the lighter materials (rocks, sand, and gravels) out the back of the sluice.

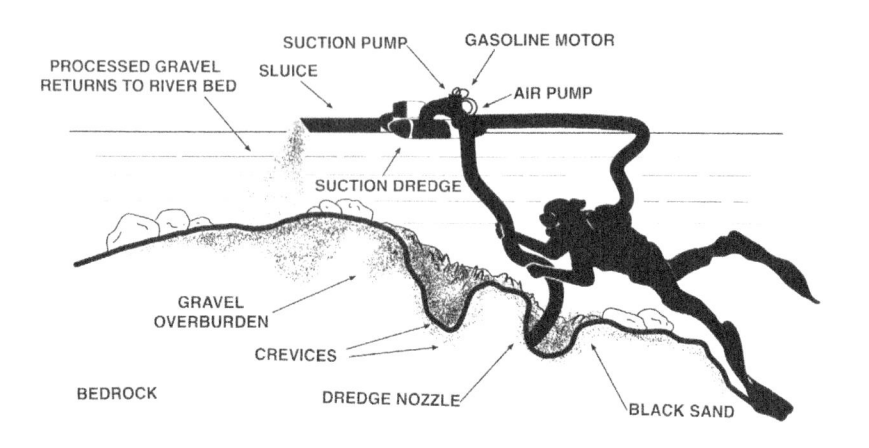

Suction dredges are rated into different sizes based on the inside diameter of the suction hose, but the end of the hose should have a suction tip that has a slightly reduced diameter to help prevent rocks and other material from clogs in the main hose. Different states will have different regulations and restrictions on how big a gold dredge can be. Common sizes include 2", 2.5", 3", 4", 5", 6" and larger, but be aware that the manufacturer generally measures the diameter of the main suction hose. The diameter of the suction tip (or nozzle) will be the limiting factor on how much material you can process, thus most dredging restrictions are based on the tip diameter, not the main hose diameter. Generally speaking, the more material you can process the better, so it is a good idea to use the largest sized dredge you can legally use in a waterway to recover the most gold.

A dredge's basic setup involves a sluice box mounted between two pontoons, with a motor and pump system and suction hose. The pump creates pressure in the suction hose to vacuum up the gravels from the stream bed. This material is then pulled up the suction hose to the surface where it is processed through a sluice box. The gravels are run over a series of riffles, lighter material is discharged out the back and the heavier materials (concentrates) are retained. At the end of the day, the miner washes out the concentrates from the sluices riffles and carpeting for further processing. At this point it has been processed down to the highest grade material, and all that needs done is to pan out the concentrates to recover the gold.

Although it may not be necessary in low water conditions, dredgers almost always use a snorkel and wetsuit so they can dive underwater and work uninterrupted. Many gold dredgers use a hookah air system, which supplies oxygen from the surface via an air compressor. This allows the modern day prospector to access deep water areas in a river that may have been inaccessible to the early day miners who didn't have this technology, thus a higher potential for good gold recovery.

Gold dredges are a bigger investment than many other types of mining equipment. A new gold dredge will generally cost between $2,000 and $6,000 depending on the size. However, the initial investment may very well pay off for a hard working gold prospector.

Highbanking

A highbanker is essentially a modern sluice box that is set up above streams or creeks and uses a water pump to pull the water up into the box so that a proper sluicing operation can take place. A highbanker is really a small scale version of larger gold mining production machinery that is used throughout the goldfields of the world. They are designed to be portable and relatively easy to set up and use.

While an ordinary sluice box can be easily hauled in a backpack, a highbanker is generally bigger, heavier and requires more planning to set up and operate properly. In fact, unless you have some additional transportation, you can count on making at least two trips to gather everything needed to set up your highbanker. Ideally, you will have vehicle access to an area so you don't have to pack heavy equipment too far.

A highbanker is basically a sluice box that uses a pump to bring water to it, making it ideal for working bench gravels that are a short distance away from the water source. Gravel is fed directly into a feed box with a grizzly to separate out large rocks. The riffles catch gold in the same way that a standard sluice box does.

While smaller, backpack size highbankers exist that can be carried by one person, they generally are not nearly as effective as their larger brethren when it comes to filtering out the materials in order to find gold. It is recommended that you get a good sized one that has good reviews and can handle a lot of use. You want a piece of equipment that can stand up to the rigors of gold mining.

Setting up the highbanker is generally no problem; just follow the instructions and make adjustments as needed. Once you get it running, you can look it over and determine what adjustments need to be made to ensure good gold retention. You can either bring a level or just eyeball the area to ensure that the proper slope is maintained.

The next step is adjusting the amount of water that is needed to properly operate your highbanker. If you don't use enough, then you will not push enough materials through the sluicing area. Too much and you can be putting gold right back into the creek or stream it came from. Ideally, you will find that gold particles get caught in the first few riffles. If you notice that a significant amount of gold is getting toward the bottom of the sluice box, then you will need to make some adjustments.

Once the operation is going, you will have to make periodic adjustments and sometimes remove larger rocks that are not being moved through. Keeping an eye on the materials flowing through is very important. You should feed the highbanker slowly so that more materials can get through the sluice without getting bogged down. This is a very common mistake that a lot of new prospectors make. Slow down and make sure that material is running through the sluice properly.

Use shorter hoses instead of one longer one to bring water up from the surface. The further up you have to move water, the harder your pump will have to work (and the more fuel you will burn).

Clay is tricky because it can actually pick up small pieces of gold that otherwise might be discovered in your highbanker. Be sure to break up clay deposits before they go through the process. This can be said for any placer mining. Overall, working with a highbanker can save you time and effort in sorting through materials. Plus, highbankers are very efficient at what they do as long as they are set up properly. If you take care of them, you will get many years of use out of them.

Trommels

A gold trommel is a piece of mining equipment that is quite popular and is used all around the world on all different sized mining operations. Its purpose is to process gold bearing material by separating out the larger rocks and boulders, allowing the smaller material that contains the gold to be run through a sluice box. Depending on the quality of your trommel and the type of material that you are feeding into it, a grizzly may also be used prior to feeding the trommel in order to separate out the largest boulders. This will allow the system to operate at its optimum potential. Some larger wash plants have grizzlies set up directly over the hopper when the material is added.

When gold bearing material is added to the hopper, it is washed down by jets of water to break up the material and help release the gold. There are different designs and amounts of water jets used, but the ultimate goal is to completely break apart any clays and mud that could retain placer gold and prevent it from being caught in the sluice.

After gold bearing gravel is added to the hopper, it enters a rotating drum that further helps to break apart the material. There are generally also water jets inside the hopper, so there is a combination of high pressure water and the rotating drum which break apart any clay and dirt as the slurry moves through the drum.

The rotating drum is constructed of screen material. This allows the smaller gravel and fines to fall through to a sluice or other processing system, while the bigger rocks will move through the drum and be discharged out the back. The ideal screen size will vary depending on the coarseness of gold found in the area, but ½" and ¾" is commonly used. If the trommel system is working properly, all of the fine material will drop through the screens to be processed, and the larger gravels and rocks will discharge out the back, completely rinsed and without holding onto any dirt or gravel.

Trommels are quite simple to build and work very well when they are constructed properly. The come in all different sizes from small units that can be used efficiently by one or two prospectors using shovels all the way up to huge commercial mining operations that use loaders to feed them.

The trommel itself does not separate out the gold. It is simply used to help separate and classify out the material for further processing. On larger operations, the trommel is generally incorporated into an entire wash plant setup, so that the smaller material that drops through the trommel is run through a simple sluice system to capture the heaviest materials and discharge the lighter sands.

To efficiently operate a trommel, it generally works best to have at least two people running the system. One miner feeds material into the trommel, while the other tends to the operation of the unit, making sure everything is working efficiently and clearing discharged waste material so more can be added.

Recovering the gold that has been processed is the same as for any other sluice box. Collect the concentrates and use some type of fine gold recovery system to complete the final process and separate the gold.

Drywashers

Most people think of creeks and rivers when they are looking for a place to go gold prospecting. The fact is that many rich gold bearing areas do not have the water necessary for panning, sluicing, or dredging. Many areas in the Southwest like Arizona, New Mexico, Nevada, and Southern California have limited water available. There are two primary ways that the small scale prospector can look for gold in these areas; metal detecting and drywashing.

Drywashers are devices that are used to separate gold from lighter material without the use of water. They do this by using a regulated air flow, which blows off lighter material and allows the gold to settle. It was actually invented by Thomas Edison in 1897, and the new tool provided miners with an invaluable way to prospect in the desert. Remember that prior to the invention of drywashers, miners needed water to process the gold nuggets out of rich gravels. This was often done by physically moving tons of gold bearing material to other areas from processing, but in many remote areas this was not feasible, and very rich areas were abandoned. With drywashers, these areas could be mined profitably.

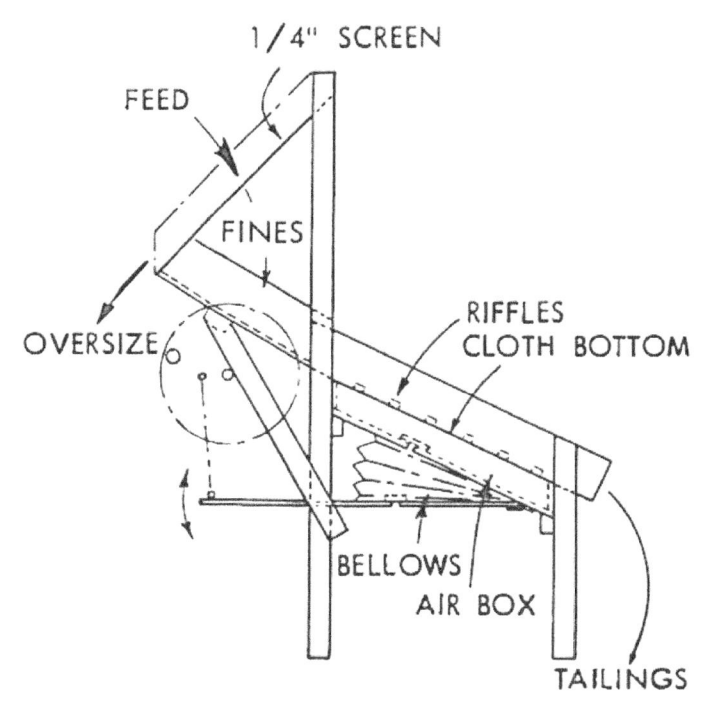

Schematic diagram of an early hand operated drywasher. Dirt is shoveled directly over a screen to separate larger rocks allowing only the finer material through. The bellows push air up from below, agitating the dirt as it vibrates over the riffles. The gold works its way down and gets caught on the upper side of the riffles.

The earliest models were hand operated, and used bellows to puff air across a series of riffles to catch the gold, while lighter material would go over the riffles and onto the ground. Many dry washers used today are powered by a small blower motor, which provides a constant flow of air that moves a counter weighted fan w hichshakes the whole apparatus, allowing gold to separate out of the lighter materials. Most have a two box setup; the upper hopper uses a grizzly to screen out the larger rocks, and the lower box has a series of riffles to capture gold. Both hand operated and motorized types are used today and each has its own benefits and drawbacks. Motor driven devices can process more material in a shorter amount of time, but the motor means that it will be louder to operate.

As with any type of prospecting, carefully searching for areas with high concentrations of gold will result in higher recoveries. Drywashers are less efficient than other methods such as a sluice box. Water will always work better than air for separation, and when possible it is always a good idea to process material by panning or sluicing rather than drywashing. However, since this is not possible in many areas, using a drywasher will often be the best option.

Drywashing works best during the summer months when the weather has been hot and dry. Due to its design, it is critically important that the material that is fed into the machine be extremely dry so that the vibration will allow it to break apart and gold to be able to separate naturally out of the mix. Many operators prefer to dig their material and let it lie out in the sun for a day or two before running it through the machine, allowing the material to dry out as much as possible. The drier it is, the more efficiently the dry washer will operate.

Another thing to consider is the use of a metal detector as a backup search tool. Despite anyone's best efforts, the design of a dry washer means that there is going to be some gold lost, either over the riffles, or rejected by the grizzly on the upper box. Using a metal detector to scan the waste piles will help ensure that you are doing the best possible recovery. A good quality VLF metal detector that can detect small pieces of gold is highly recommended.

Metal Detecting

Finding gold nuggets with a metal detector may seem easy, but it can in fact be one of the most challenging types of prospecting there is. It certainly seems simple enough; you swing your detector until the coil is over a gold nugget, the detector sounds off, and you dig it up. In reality, while it can be very productive, using a metal detector for gold prospecting can also be extremely frustrating, so let's discuss a few of the challenges that you are likely to face when you start looking for gold with a metal detector, and a few tips that will help you get started on the right track.

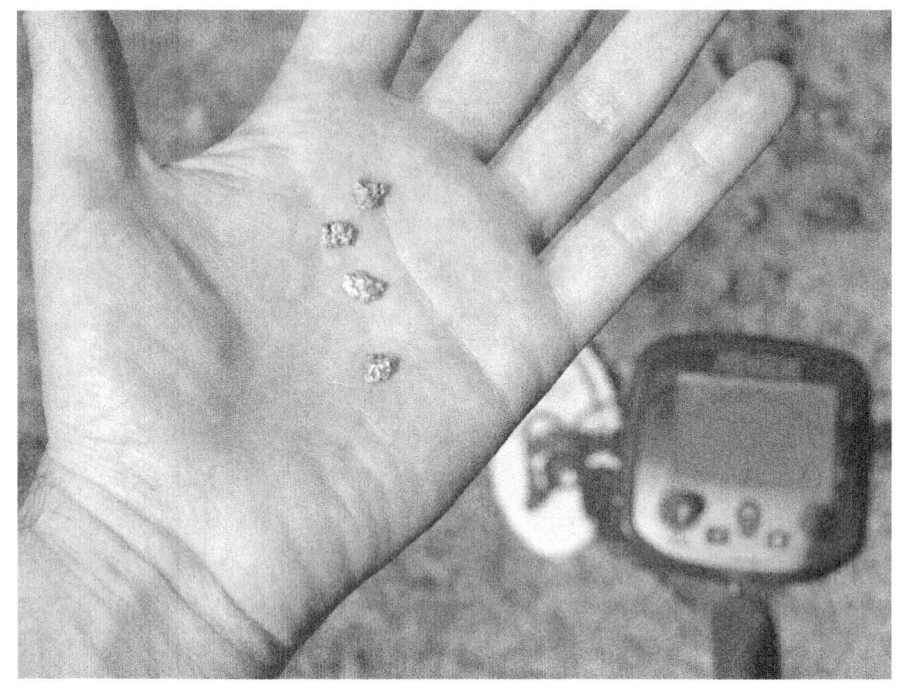

Four nice gold nuggets found in Northern California. These were found with a Fisher Gold Bug Pro metal detector.

The first thing worth emphasizing is to use a metal detector that is specifically designed for finding gold. Just about all metal detectors can detect gold, and manufacturers will often claim that their detectors are good at finding gold, but the truth is only a select group is specifically designed for nugget shooting. The average detector that might do just fine locating coins in a park will probably struggle if you take it out to the gold fields. Areas that produce gold nuggets often have a unique set of challenges associated with them, so let's discuss each of them in detail.

One of the biggest problems with most gold bearing areas is the highly mineralized ground that is associated with it. Most ground contains varying amounts of iron, and gold is usually found in areas that have extremely high amounts of it. This causes the vast majority of metal detectors to struggle, as they will sound off constantly because they are sensing the minerals in the ground.

The other pest that you will find in many areas is hot rocks, which are rocks that contain a high amount of iron that will make your detector sound off as well. A detector with ground balancing is absolutely essential, but still many detectors will have trouble with this, and it can be just about impossible to cancel out the chatter produced by the mineralized ground to be able to distinguish the difference between it and a gold nugget.

Another problem that is common in many gold areas is high amounts of iron trash. The miners in the early days didn't think much about leaving behind junk. Often they were camped right on their claims, leaving behind old cans, boot tacks, nails, bullets, shovels, snuff cans, and just about anything else you can imagine. High amounts of trash can sometimes be overwhelming, even to an experienced detectorist, but it is something that has to be dealt with if you want to be a good nugget shooter.

Another thing that makes metal detecting for gold nuggets a real challenge is that the vast majority of them are very small. While we would all love to dig up those nice softball sized nuggets that we dream about, the reality is that most of the nuggets that are found with a metal detector are pretty small. We are talking little flakes that might be smaller than a grain of rice. So in order to successfully find gold nuggets on a regular basis we need a detector that is sensitive enough to find the small nuggets, while at the same time be able to handle highly mineralized ground and also distinguish the difference between ferrous trash and a gold nugget.

So now that we have outlined the biggest challenges that we will have to deal with when looking for nuggets, let's talk specifically about which detectors are best at dealing with these challenges. I will go ahead and outline some specific brands and models that have good reputations as "gold getters," but everyone has their own opinions about what is best. Do your research, weigh the options, and figure out which is best for your needs.

You don't need to buy the most expensive detector on the market. Many successful nugget shooters are using detectors that are decades old and still find plenty of gold. The skill of the operator is more important than the specific detector.

Without getting too deep into specifics, understand that there are basically three types of technology used in metal detectors today; Very Low Frequency (VLF), Pulse Induction (PI), and Zero Voltage Transmission(ZVT).

VLF metal detectors are an older technology, but are used in the majority of the detectors on the market today. They are best at locating small nuggets at fairly shallow depths. PI metal detectors are a newer technology and were specifically designed to detect larger nuggets at deeper depths. While they excel at searching deeper into the soil, they will miss very small nuggets that a VLF will find. The latest technology being used in some metal detectors is ZVT, which does quite well on smaller gold at depth.

In no particular order, some of the most popular VLF metal detectors are the Fisher Gold Bug 2 and Gold Bug Pro, Garrett AT Gold and Goldmaster 24k, Minelab Gold Monster and Equinox 800. All of these detectors are well respected for their ability to find small nuggets, as well as discriminate trash and handle soil mineralization fairly well.

Minelab is by far the winner when it comes to the latest metal detector technology. With a few exceptions, they have been the only show in town when it comes to high quality PI detectors, and they are by far the favorite of most serious nugget shooters. The older models include the SD2100, SD 2200 and GP3000, GPX4500 and GPX 5000. The Whites TDI and Garrett ATX are two PI detectors, but currently don't have near the fan following that Minelab has.

The Minelab GPZ7000 and GPX6000 are the latest offerings as I am writing this book. Both are excellent detectors, with a high price tag to match.

The very first thing you should do when you get a new metal detector is read the owner's manual. After you have read it, start at the beginning and read it again. The key to successfully finding gold with a metal detector is understanding your machine, so take the time to really learn it – how it works, the different adjustments, how to ground balance it, etc. One thing all successful nugget shooters have in common is an intimate knowledge and understanding of their machine.

If there is one "secret" to finding gold, this is it; search where gold nuggets have been found before. Some areas have produced plenty of gold, but not in sizes large enough to be found with a metal detector. Many of the goldfields throughout the US would fall into this category. Some areas have produced gold, but only fine dust or tiny micron gold. If you want to find gold with a metal detector, you need to seek out areas that have a history of producing nuggets large enough to be heard by your detector.

Don't just metal detect randomly. A common mistake is to assume that just because gold was found in a general area, that you can go anywhere in that area, turn on your metal detector and start digging up nuggets. Find the exact areas that the old timers worked. Look for old tailing piles, places where miners' hands stacked rocks along a creek. Find those prospects in the side of the hill where someone did some digging. Look for large areas dug up with bucket line dredges or big hydraulic pits. If you are in an arid region, look for drywasher piles left behind by the old timers. Anything that shows sign that gold has been found in the area before is a good indication that it can be found today. The early prospectors were great at finding gold, but they didn't have metal detectors, so you have the chance of finding gold that they missed.

Once you have found a good looking area, it's time to start metal detecting. The first thing you need to do is ground balance your metal detector to the mineralization in the soil. Since you read your owner's manual you already know how to do this (you did read it, right?) If you really have found a good location that was worked by the early day gold prospectors, there is a good chance that right away you're going to get a strong signal from your detector, and you will dig a rusty nail. You'll move forward a few feet and get another hit, and you will dig up another rusty nail. You will go another few feet and dig up ten more nails, an old beer can, four little unidentified bits of iron, and a piece of bird shot. You have learned about the scourge of the modern day prospector… trash.

This is where learning the "language" of your metal detector will really pay off. The fact is that even guys that are really good at finding gold nuggets with a metal detector still dig a lot of trash. It is just a part of the hobby, and it is something that you just have to deal with to some extent. However, learning your machine will really help cut down on the amount of digging you have to do.

Quality discrimination can separate out much of the nonferrous targets that you go over with your detector, either by producing a different tone than gold makes, or by blanking them out completely. If you rely too heavily on your detector's discrimination feature, you will miss gold. A lot of detectorists dig all targets, regardless of the tones or numbers that the metal detector gives them. Digging every target is the only way to be 100% sure that you aren't walking past a nugget.

It's always a good idea to bring a small test nugget out into the field with you. Take a small nugget and glue it to a poker chip or guitar pick, something that is not made of metal and large enough that you won't lose it. This will help you tune your ear to the sound of gold and what you should be listening for.

Here's another bit of encouragement. If you are digging small pieces of lead, you are on the right track. It is almost impossible to distinguish the difference between lead and gold. While this can definitely be frustrating if you get into an area that has bullets and bird shot everywhere, you should also think of it as a good thing. If you are able to detect a small piece of bird shot, you will be able to find a small piece of gold. Often it just takes time, and maybe hundreds of pieces of trash before you are lucky enough to find that first piece of gold, but if you are persistent you will be rewarded.

Metal detecting for gold nuggets is the toughest type of detecting there is, certainly harder than digging up coins at the park. Nuggets are generally small, the ground is highly mineralized and noisy, and often there is metal rubbish scattered everywhere. Still, when you find that first gold nugget you might just be hooked for life.

Cleaning and Selling Your Gold

Getting Fine Gold Out of Black Sand

Recovering fine gold from black sands is a challenging final step in gold recovery. The vast majority of gold that we find as prospectors is very small in size, ranging from gold flakes down to dust so small that it is nearly invisible. Extremely fine particles of gold like this can be very difficult to separate from black sands and other heavy minerals that have little to no value. Despite their small size, the accumulation of this fine gold can add up to substantial amounts, so it is in the prospector's best interest to find a process that will help aid in the recovery of this fine gold to maximize profits.

The process required to separate gold from lighter minerals is simple enough to understand. Gold, being the heaviest element we typically encounter in a stream, is easily separated from lighter materials by various gravity separation methods. The difficulty lies in the fact that even the best tuned mining equipment will retain heavy minerals in addition to gold. The end result of a day of gold mining will result in what we call "concentrates." If you are in a gold bearing area, there is a good likelihood that you have captured some gold, but along with that gold you are going to find black sands and some other trace minerals.

First, let's take a look at what concentrates are generally comprised of. If your mining equipment is set up properly, then you should be discharging the vast majority of the lighter material, such as most light sands and gravel. Black sands, with a specific gravity of around 5 (meaning an equivalent volume of black sands is five times heavier than water), is generally made up of hematite and magnetite. Magnetite is known to be highly magnetic, and this will help in the separation process. Hematite on the other hand is only slightly magnetic. It does not hold near the attraction to a magnet that magnetite does, but will still draw to a magnet if it is extremely strong. This will be to the miner's advantage further in the separation process. Other heavy minerals that are commonly encountered can include cinnabar, platinum, diamonds, lead, mercury, garnets, and other assorted heavy minerals. Generally if present, these will only comprise a small percentage of the total concentrates. Black sands are the main challenge that prospectors will have to contend with.

So where do all these black sands come from anyways? Both magnetite and hematite are crystalline oxides of iron, which is one of the most common elements on the planet. In addition, it is a common fact that metamorphic rock and igneous rock are generally associated with iron deposits as well as gold deposits. While there are occasionally gold bearing areas that have very little black sands, it is generally accepted that there will be at least some amount of black sands anywhere that gold is found in varying quantities. In some areas you will find just a trace, but most areas will produce a significant amount, thus your concentrates will have a high percentage of black sands.

There are many different methods that a gold prospector can use to separate the gold out of their black sands. These methods range from as simple as carefully gold panning the concentrates several times, to expensive separation equipment costing thousands of dollars. For most small to medium scale prospectors, the more expensive separation methods such as shaker tables are not the best method. These can be very valuable for a large mining operation, but the average prospector may never recover enough additional gold to justify such a large purchase. On the other hand, simple hand panning of concentrates will result in lost gold, especially the finest material. No matter how skilled a gold panner is, it is impossible to work as efficiently as a mechanized piece of equipment specifically designed for fine gold separation.

Fortunately there are quite a few good fine gold recovery systems on the market that are both efficient and cost effective. A few hundred dollars will generally get a prospector set up with a system that will do an excellent job of retaining the finest gold in their concentrates.

A favorite among prospectors is the blue bowl. It uses a simple design of circulating water (much like a flushing toilet) to separate any lighter material and black sands while keeping the gold at the bottom of the bowl. Other popular equipment include spiral panning wheels, which use a specially designed spiral motion at an exact angle to capture the fine gold. Micro sluices work much the same as a standard sluice, but using much smaller riffles and less water flow.

All of these types of equipment work, and can be purchased for a few hundred dollars or less. The trick to maximizing the efficiency of any of these methods is to classify your final concentrates down. Reducing down to approximately -30 mesh will help reduce losses. Anything larger than -30 mesh should be panned out and inspected for larger pieces of gold. Once classified, the black sands can be run through the equipment and yields will be maximized.

Although a bit more expensive than other methods, the Gold Cube is a very popular tool for fine gold recovery that is used by many gold prospectors.

Even after final processing, you will still find that your concentrates, which are now "super concentrates", will still have a mix of fine gold and fine black sand particles. Here is where the magnetic nature of black sands will help us in the final separation process. Using a rare earth magnet, we can physically pull them away from the gold. It is very important to get an extremely powerful rare earth magnet, because although magnetite is highly magnetic, hematite is only slightly magnetic. The added power of one of these extra strong magnets is the key to this process.

Place the remaining concentrates in the bottom of a plastic gold pan with an inch of water. Spread them out across the bottom of the pan, and hover your magnet just above them. The goal here is to actually cause the black sands to "jump" out of your gold pan. Clean the accumulated black sands from the bottom of your magnet and repeat this process several times until the black sands are nearly all removed. Be sure to inspect the black sands as you clean them off the bottom of your magnet, as small bits of gold can be trapped and lost in this part of the process. Once you have removed nearly everything but the gold, use your magnet and stir the material at the bottom of the pan to pick out the remaining fine textured sands.

Once finished with this wet separation process, you should be down to almost clean gold. Dry whatever you have left and inspect it carefully. There will likely still be a few particles of mineral other than gold, but they should be reduced down to a small enough quantity that they can carefully be separated out with the tip of a sharp knife, or carefully blowing on the remaining material. It takes time and patience.

In the past, mercury was commonly used for final cleanup, but with our current knowledge about the dangers, plus the invention of these excellent fine gold separation tools, mercury really isn't needed for gold mining like it once was. Using the process described above will work extremely well for the needs of most gold prospectors.

As mentioned earlier, large mining operations may benefit from more professional equipment designed to process large amounts of black sand. However, we are talking about mines that produce several tons of concentrates, rather than a few five-gallon buckets full like most prospectors.

It might also be noted that some prospectors seem to get obsessed with retaining fine gold. They almost have a delusional belief that there are ounces and ounces of gold hidden in their concentrates, and the perfect extraction method will somehow produce more and more gold. The truth is that there is only so much gold in your concentrates, and no perfect method is going to produce more gold than there actually is. Additionally, spending an extensive amount of time capturing micro gold particles can become a bit pointless, since a visible speck of gold may only have a value of a few cents. Do your best to retain the finest gold particles that you can, but don't spend so much time that you take away from time that would be better spent out in the field. Using the process outlined above will do an excellent job of fine gold recovery.

Separating Gold by Size

Once you've got a nice amount of placer gold, you should separate out your larger nuggets and "pickers" from your fine gold dust, since you can often get a nice premium for select pieces. The easiest way to do this is to use classifiers with different mesh sizes.

Classifiers can be purchased from mining supply shops. They stack together and make quick work of sorting your gold into different screen sizes. Stack them on top of each other, with the coarsest screen on top and finest screen on the bottom. Pour your vial of gold onto the classifiers and gently agitate them. Make sure you have a gold pan underneath your finest classifier on the bottom, so you will catch the smallest specks of gold dust that fall all the way through.

Gold in Rock - Crushing Ore

For those prospectors out there who have dug up hard rock ore samples and want to test them for their gold content, there are a few different options that are available to you. But for many independent and smaller operations, you may realistically be limited on which type of testing you can use based on its overall costs. For most small-scale miners, you will be most interested in the free gold content within your ores. This is the gold that can be obtained without the use of more complex and highly regulated methods.

Most lode mining in the U.S. is done by commercial mining companies on a large scale, primarily because of the high cost involved with startup and operation. This means that for companies with substantial resources, they have more options open to them than the smaller operators or individual miners. Mines that are expecting to process millions of pounds of ore can cost effectively consider more expensive extraction methods due to the huge amounts of material that they are processing.

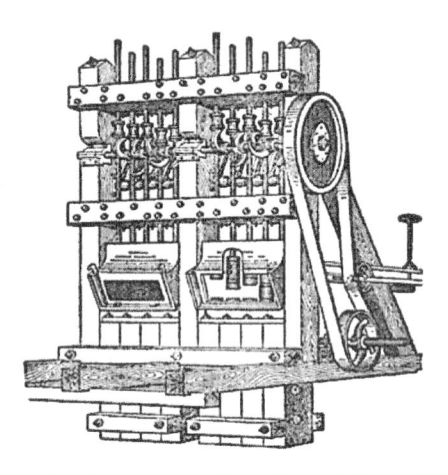

Stamp mills are great for large operations, but a bit overkill for the average prospector.

Probably the easiest way to test the gold values in your ore sample is to have an assay test performed by a local, independent assay firm. This is the simplest way to test your hard rock ore samples, but it is also one of the most expensive as well. Plus, the total gold content within your ore is only valuable to you if you have an extraction method that can collect it. Many ores contain gold that is locked up in sulfides, and cannot be recovered without significant expense that is out of reach for many miners.

An alternative to an assay test is to determine just the free gold content found in the ore. Free gold is the portion that is easily removed without the use of complex chemical procedures. Miners have been extracting free gold from hard rock sources for thousands of years. It simply requires that the ore be crushed and broken apart enough that the gold can be separated from the host material. Arrastras and stamp mills were commonly used at mines all throughout the world for this purpose. On an even smaller scale, prospectors can crush up pieces of ore using a mortar and pestle, or "dolly pot", which is just a steel pipe that ore is placed in and crushed up. Reasonably priced rock crushers can also be bought that will make short work of any specimens that you want to break up.

A heavy-duty mortar & pestle works well for crushing small samples of gold-bearing quartz. Use to pulverize ores into fine dust, then pan the material to separate out the gold.

Once the ore has been broken up into fine dust, testing for free gold content in hard rock ore can be done using some type of gravity method. A gold pan can be used to sample small amounts of crushed ore, or if a decent water supply is available then it can be run through many types of separation devices. Shaker tables are commonly used, and although quite expensive, work exceedingly well for recovering small particles of free gold.

The traditional chemical treatments done by large mining companies uses cyanide, which has a very significant permitting process that is unattainable for the average prospector here in the United States. For most miners, some type of gravity system is the best choice, as it does not involve any chemicals and does not require the same level of permitting (or none at all).

Regulations have basically squeezed the smaller operators of hard rock mines out of the picture. Smelting plants that used to purchase considerable amounts of hard rock ore from prospectors have mostly gone away due to the high regulatory costs. While the high cost of extraction may limit the values that can be obtained from certain material, ores that have a considerable amount of free gold may still be mined profitably by the average miner.

Once an ore has been crushed and the gold extracted, evaluations can be made on the overall value or potential for a mining operation. Mining for lode gold has a very high failure rate, mainly due to the extremely high operating costs that it takes to recover lode gold. Additionally, it is important to realize that a few small samples of ore may not be representative of a large area. Before undertaking any lode mining operation, a large amount of sampling should be done. There are many miners who have lost a lot of money by testing a few high grade pieces of ore and assuming that all the ore in the mine would be just as rich.

Methods of Cleaning Gold

If you are lucky enough to find a nice nugget, you may want to clean it up a bit. Sometimes nuggets that come out of rivers are bright and polished requiring no cleaning at all, but nuggets that are dug out of the ground are often caked with various different muck, dirt, caliche, ironstone, and staining which will hide the beauty of the gold hidden underneath. On these nuggets, some type of cleaning will often enhance the nugget, both in its visual appeal and with its overall value to collectors as well.

I have heard dozens of different techniques that prospectors have used, and all of them work to a certain degree. Most involve soaking in some type of household cleaner. My recommendation would be to use the gentlest method that will get the job done.

Any recently dug nugget is probably going to have some dirt and grime stuck to it. Soaking the nugget for 20 minutes in some hot soapy water may be all that is needed to break up some of the grime that clings to them. If you remove the nugget from the hot water and you see some improvement, gently scrub it with a toothbrush and put it back in the water to soak for a few more minutes. Repeating this process may be all that is needed to clean up some nuggets for display.

An ultrasonic cleaner will also work wonders to break up the grime that is stuck to nuggets. These are typically used to clean jewelry, and can be purchased for a very reasonable cost. Adding a jewelry cleaning solution can help with the process. There is also a cleaning product called Simple Green that does an excellent job when diluted with water in the ultrasonic cleanser. I have found that this method will work for the majority of dirty gold nuggets, and it does a fine job of removing the grime without giving the nuggets an unnaturally clean look.

Another great way to clean up dirty nuggets is to use vinegar and salt. Using a bottle with a tight fitting lid, fill it part way with vinegar and add enough salt that you can see the salt accumulate at the bottom of the vial without dissolving. Drop a few nuggets in the vial and gently shake it for a few minutes. The rough salt works as a slight abrasive to scrub the nuggets, and the vinegar helps to break up some of the grime and stains. Depending on how much cleaning is needed, you can leave the nuggets in the bottle for a few days or even a couple weeks, giving the mixture a gentle shake every once in a while. You will notice that the vinegar will take on a stained color, indicating that the grime is being removed from the nugget.

Sometimes hard caliche and ironstone material will be very stubborn, and can be difficult to remove from nuggets. This is where some type of cleaner or mild acid might be needed to clean up the nuggets. Most rust removers that are in the household section of any department store will do the job. One product that was recommended to me several years ago is a rust remover called Whink Rust Remover. It does a fine job of breaking up tough iron deposits. CLR is another well-known product that can get the job done on some of the more stubborn nuggets. It's a good idea to wear protective gloves and goggles if you intend to use any of these products.

I generally try to avoid using harsh chemicals unless necessary, as I find that they can take away some of the natural look of the nugget by actually making the nuggets look too shiny. This is why I would recommend using the gentle soap water scrub if possible, and hold off on using any chemicals until you have determined whether or not they are really necessary.

A method that is used by some prospectors to clean their nuggets is soaking them in hydrofluoric acid. This is an extremely dangerous acid that will remove any and all material that is attached to a nugget including quartz rock. Often high-grade specimens with intricate crystalline gold are etched with acid to expose the gold. The most common use for this is on gold in quartz specimens where the gold is contained within quartz and needs to be completely removed to expose the gold. Hydrofluoric acid is extremely dangerous; in fact just a few drops can cause serious burns and can even be fatal if it comes in direct contact with the skin. I personally will not get near the stuff, and would not recommend it as a way to clean nuggets. There are much safer ways to clean nuggets as previously mentioned in this article, and hydrofluoric acid is simply unneccessary for most jobs. If you have a specimen that you are dead set on using acid to clean, I would strongly recommend finding a professional to do the job. Search the internet for pictures of hydrofluoric acid burns and you will see why I recommend avoiding it.

Fair Prices Depending on Size and Quality

There are a few different ways to market the gold that you find. Getting the best price will depend on a variety of factors, but I have noticed that by far the *worst time* to sell gold is when you need to sell it quickly!

With any gold that you find, it's important to remember that it will NEVER be 100% pure gold (24k). Natural gold from the Earth is alloyed with other metals, primarily silver and copper. There are many areas around the world where natural gold can be 23k and higher, but it more commonly ranges closer to 20k to 22k. Some mining districts commonly produce gold closer to 18k. I have seen nuggets that were as low as 16k.

Selling Gold Dust, Nuggets and Specimens

Refineries - Obviously, the lower the gold content, the less you can expect for your gold. This is especially true if you send your gold into a refinery. Refineries don't care about how pretty your gold is, they are only interested in the gold content. They will analyze your gold and determine its purity. From there, they take a % out as profit, and offer you a price based on that.

When selling your gold to refineries, you can usually expect to get anywhere from 70% to 85% of the spot price. This may not be the best option if you want to get top-dollar, but it's probably the easiest option.

Collectors, Friends and Family - Selling to a collector is an option that may be worth considering. However, small gold dust and flakes of placer gold are usually not as collectible as gold nuggets are, so finding a buyer that will pay a premium may be difficult. However, if you aren't in a rush, this may be a viable option, especially if you have smaller quantities of just an ounce or two.

Selling small amounts to friends and family is a good option for some people.

I have found that most people these days have never even seen gold in its natural form, and they are quite fascinated when they see a piece of raw gold. You can usually get a nice premium from these folks if you sell them a piece or two, since they are less interested in the "gold value," and more interested in owning a piece as a souvenir.

Selling gold online is an option. Some miners sell their gold on eBay and do relatively well, although the fees and shipping costs can often reduce the profits. You'll also have to deal with the obvious risks of selling through the mail.

Large Quantities - The quantity of placer gold that you are trying to sell is another consideration. If you just have a few ounces, you can take the time to search for a collector who might pay a little extra for it. However, if you are mining several ounces of gold per week in a larger operation, you will find it difficult to find a buyer who will pay a premium AND buy large amounts. Getting a premium price generally only happens in small quantities.

Making Jewelry - Probably the best way to get a premium price for your gold is to use it to make jewelry and sell. You can use small nuggets to make earrings, rings and pendants that you can then sell for several times the spot price of the metal. Of course, learning how to become a goldsmith is beyond the scope of this book, but it is certainly an option that you might want to consider. I took a few Community Education classes at my local arts college and learned some basic soldering skills that have helped me market some of my nicer nuggets. Just remember, only a very small percentage of the gold you find will be considered "jewelry grade."

Printed in Great Britain
by Amazon